Study Guide to Accompany

Managerial Accounting
for the
Hospitality Industry

Lea R. Dopson

David K. Hayes

JOHN WILEY & SONS, INC.

This book is printed on acid-free paper. ∞

Copyright © 2009 by John Wiley & Sons, Inc. All rights reserved

Published by John Wiley & Sons, Inc., Hoboken, New Jersey.

Published simultaneously in Canada.

No part of this publication may be reproduced, stored in a retrieval system, or transmitted in any form or by any means, electronic, mechanical, photocopying, recording, scanning, or otherwise, except as permitted under Section 107 or 108 of the 1976 United States Copyright Act, without either the prior written permission of the Publisher, or authorization through payment of the appropriate per-copy fee to the Copyright Clearance Center, Inc., 222 Rosewood Drive, Danvers, MA 01923, 978-750-8400, fax 978-646-8600, or on the web at www.copyright.com. Requests to the Publisher for permission should be addressed to the Permissions Department, John Wiley and Sons, Inc., 111 River Street, Hoboken, NJ 07030, 201-748-6011, fax 201-748-6008, or online at http://www.wiley.com/go/permissions.

Limit of Liability/Disclaimer of Warranty: While the publisher and author have used their best efforts in preparing this book, they make no representations or warranties with respect to the accuracy or completeness of the contents of this book and specifically disclaim any implied warranties of merchantability or fitness for a particular purpose. No warranty may be created or extended by sales representatives or written sales materials. The advice and strategies contained herein may not be suitable for your situation. You should consult with a professional where appropriate. Neither the publisher nor author shall be liable for any loss of profit or any other commercial damages, including but not limited to special, incidental, consequential, or other damages.

For general information on our other products and services, or technical support, please contact our Customer Care Department within the United States at 800-762-2974, outside the United States at 317-572-3993 or fax 317-572-4002.

Wiley also publishes its books in a variety of electronic formats. Some content that appears in print may not be available in electronic books. For more information about Wiley products, visit our Web site at http://www.wiley.com.

Library of Congress Cataloging-in-Publication Data:

ISBN: 978-0-470-14055-0

Printed in the United States of America

10 9 8 7 6 5 4 3 2 1

Table of Contents

	To the Student	v
PART I:	**ACCOUNTING FUNDAMENTALS**	
Chapter 1	Hospitality Industry Accounting	1
Chapter 2	Accounting Fundamentals Review	10
PART II:	**FINANCIAL STATEMENTS**	
Chapter 3	The Income Statement	26
Chapter 4	The Balance Sheet	46
Chapter 5	The Statement of Cash Flows	60
Chapter 6	Ratio Analysis	72
PART III:	**MANAGEMENT OF REVENUE AND EXPENSE**	
Chapter 7	Food and Beverage Pricing	98
Chapter 8	Revenue Management for Hotels	114
Chapter 9	Managerial Accounting for Costs	132
PART IV:	**ACCOUNTING INFORMATION FOR PLANNING**	
Chapter 10	Forecasting in the Hospitality Industry	148
Chapter 11	Budgeting and Internal Controls	160
Chapter 12	Capital Investment, Leasing, and Taxation	180
Answers to Key Terms & Concepts Review, Discussion Questions and Quiz Yourself		199

To the Student

Managerial Accounting for the Hospitality Industry is a book that we the authors were particularly excited and privileged to develop. We view it as an essential book not merely because the topic of managerial accounting is so important, but more critically, because the success of hospitality students themselves is so important. In addition to their "people skills," all hospitality professionals must possess "number skills." Hospitality is a business, and it has been said that accounting is the language of business. A clear understanding of that language simply must be acquired by all who wish to achieve their business's financial and profit objectives. It is the goal of this book to provide each student with that crucial understanding.

Learning managerial accounting by reading this ***Study Guide*** will be fun. That's a promise from the authors to you. It is an easy promise to keep because working in the hospitality industry is fun. And it is challenging. Learning managerial accounting (accounting used by managers) will be exactly the same: fun and challenging. In this ***Study Guide***, you will see how hospitality managers actually consider and apply accounting information.

If you work hard and do your best, you will find you do have the ability to master all of the information in this ***Study Guide***. When you do, you will have gained an invaluable management tool that will enhance your company's performance and, by doing so, advance your own hospitality career.

This ***Study Guide*** is organized according to the 12 chapters in the text. Each chapter provides the following student aids:

- Highlights
- Study Notes
- Key Terms & Concepts Review
- Discussion Questions
- Quiz Yourself

The "Study Notes" are intended to help you better understand the content. The answers to odd-numbered "Key Terms & Concepts Review," "Discussion Questions," and "Quiz Yourself" questions are provided in the back of the workbook. All the information in this ***Study Guide*** can be used for reviewing the material and for testing your grasp of hospitality managerial accounting concepts and techniques.

Special thanks go to Nancy Kniatt who assisted us with this ***Study Guide***. We hope that this exposure to the study of hospitality managerial accounting creates in you the same enjoyment that we have experienced in our careers. If so, everyone will have found it a rewarding experience.

Lea R. Dopson
David K. Hayes

Part I: Accounting Fundamentals

Chapter 1

Hospitality Industry Accounting

Highlights

* Unique Aspects of the Hospitality Industry
* The Purpose of Accounting in the Hospitality Industry
* Branches of Accounting
* Why Hospitality Managers Use Managerial Accounting
* The Uniform System of Accounts
* Ethics and Hospitality Accounting
* Key Terms and Concepts Review
* Discussion Questions
* Quiz Yourself

Study Notes

Unique Aspects of the Hospitality Industry

- **Hospitality** can be defined as the friendly and charitable reception and entertainment of guests or strangers.
- While each of the industry sub-segments are very different and can be classified in very different ways (for example, profit vs. non profit; or corporate vs. privately owned), one way to classify them is by their emphasis on either lodging or food and beverage (F&B) services. It is the emphasis on providing lodging and meals (in a variety of settings) that distinguishes those who are considered to be working in the "hospitality" industry.
- Within the lodging and food services industries are a variety of related fields, including hotels, restaurants, clubs, resorts, casinos, cruise ships, theme parks; the recreation and leisure market: arenas, stadiums, amphitheaters, civic centers, and other recreational facilities; the convention center market; the education market: colleges, universities, and elementary and secondary school nutrition programs; the business dining market: corporate cafeterias, office complexes, and manufacturing plants; the health care market: long-term care facilities and hospitals; and the corrections market: juvenile detention centers and prisons.

- The number of opportunities offered by the hospitality industry is significant, as are the opportunities for those managers who understand and can utilize their hospitality accounting skills.

The Purpose of Accounting in the Hospitality Industry

- **Accounting** is the process of recording financial transactions, summarizing them, and then accurately reporting them. An **accountant** is a person skilled in the recording and reporting of financial transactions.
- Accounting is utilized by all managers in business. Accounting in the hospitality industry is utilized every time a guest purchases food, beverages, or a hotel guest room.
- Businesspersons estimate their costs before they decide to build hospitality facilities and often seek loans from banks to help them. Banks want to know about the proposed business's estimated financial performance before they decide to lend it money.
- The owners of a hospitality facility want to monitor their business's financial condition; investors want to put their money in businesses that will conserve or increase their wealth. To monitor their investments, owners and investors seek out and rely upon accurate financial information.
- Accounting is not the same as management; accounting is a tool used by good managers.
- Hospitality managers must learn to use accounting techniques as well as their education, experience, values, and goals to make the very best management decisions possible for themselves and the businesses they are responsible for managing.

Branches of Accounting

Financial Accounting

- Business accountants who specialize in **financial accounting** are skilled at recording, summarizing, and reporting financial transactions. Financial transactions include **revenue**, the term used to indicate the money you take in, **expense**, the cost of the items required to operate the business, and **profit**, the dollars that remain after all expenses have been paid.

$$\text{Revenue} - \text{Expenses} = \text{Profit}$$

- Financial accounting also includes accounting for **assets**, which are those items owned by the business; **liabilities**, which are the amounts the business owes to others; and **owners' equity**, which is the residual claims owners have on their assets, or the amount left over in a business after subtracting its liabilities from its assets. These transactions can be used to develop the following equation for the balance sheet:

> **Assets = Liabilities + Owners' Equity**

Cost Accounting

- **Cost accounting** is concerned with the classification, recording and reporting of business expenses.
- For cost accountants, a **cost**, or expense, is most often defined as "time or resources expended by the business."
- Cost accountants determine costs by departments, by business function or area of responsibility, and by the products and services sold by the business.

Tax Accounting

- A **tax** is simply a charge levied by a governmental unit on income, consumption, wealth, or other basis.
- **Tax accounting** concerns itself with the proper and timely filing of tax payments, forms, or other required documents with the governmental units that assess taxes.
- Professional tax accounting techniques and practices ensure that businesses properly fulfill their legitimate tax obligations.
- Some of the taxes hospitality managers may be responsible for include **occupancy taxes**, **sales taxes** and **payroll taxes**.

Auditing

- The auditing branch of accounting is chiefly concerned with the accuracy and truthfulness of financial reports. It is designed to point out accounting weaknesses and irregularities and thus prevent accounting fraud.
- An **audit** is an independent verification of financial records. An **auditor** is the individual or group of individuals that completes the verification.
- The auditing branch is also concerned with safeguarding the assets of a business from those unscrupulous individuals who would seek to defraud or otherwise take advantage of it.
- The total collapse of the Enron Corporation in late 2001, as well as other highly publicized business failures, demonstrates the importance of auditing. Rampant violation of standardized accounting rules led Enron's investors, creditors, employees, and others to believe the company was financially sound when, in fact, it was not.
- In part because of the potential damage that could be done by unscrupulous corporate managers, in 2002 the United States Congress passed the **Sarbanes-Oxley Act (SOX)**. This law provides criminal penalties for those found to have committed accounting fraud, and also covers the regulation of auditors assigned the task of verifying a company's financial health.
- Individuals who are directly employed by a company to examine that company's own accounting procedures are called **internal auditors**. **External auditors** are

individuals or firms who are hired specifically to give an independent (external) assessment of a company's compliance with standardized accounting practices.
- In larger hotels, the **controller**, who is the person responsible for managing the hotel's accounting processes, may serve as the auditor. In very large properties, full-time individuals are employed specifically to act as the property's in-house auditors.

Managerial Accounting

- **Managerial accounting** is the system of recording and analyzing transactions for the purpose of making management decisions.
- Its proper use requires skill, insight, experience, and intuition. These are the same characteristics possessed by the best hospitality managers.
- A brief summary of the branches of accounting and the main purpose of each can are shown in Figure 1.1.
- In the United States, those individuals recognized as highly competent and professional in one or more of the branches of accounting have earned the designation of **Certified Public Accountant (CPA)**. To become a CPA, a person must meet the requirements of the state or jurisdiction in which they want to practice.
- Hospitality professionals who work extensively in the areas of accounting and technology often become members of the **Hospitality Financial and Technology Professionals (HFTP)**. HFTP offers its own certifications for hospitality professionals working in the accounting and technology areas, and provides a global network for them.

Why Hospitality Managers Use Managerial Accounting

- **Hospitality accounting** is not a separate branch of accounting, but it is a very specialized area that focuses on those accounting techniques and practices used in restaurants, hotels, clubs, and other hospitality businesses.
- Those practicing managerial accounting in the hospitality industry have specialized knowledge. That knowledge is the result of learning the intricacies of the restaurant or hotel business and then applying what they know to a financial analysis process.
- Refer to Figure 1.2 for a hospitality accounting term quiz which will confirm some of the reasons why managerial accounting is a separate field of study.

Uniform System of Accounts

- Many hospitality companies require that their managers use a series of suggested (uniform) accounting procedures created specifically for their own segment of the hospitality industry.
- A **uniform system of accounts** simply represents agreed upon methods of recording financial transactions within a specific industry segment.

- In the hospitality industry, some of the best known of these uniform systems are:
  - **Uniform System of Accounts for Restaurants (USAR),** developed for the restaurant industry by the National Restaurant Association (NRA).
  - **Uniform System of Accounts for the Lodging Industry (USALI),** developed for the lodging industry by the Hospitality Financial and Technology Professionals (HFTP) and the Educational Institute (EI) of the American Hotel & Lodging Association (AH&LA).
  - **Uniform System of Financial Reporting for Clubs (USFRC),** a club accounting resource for club managers, officers, and controllers produced through the joint efforts of Hospitality Financial and Technology Professionals (HFTP) and the Club Managers Association of America (CMAA).

Ethics and Hospitality Accounting

- Sometimes it may not be clear whether an actual course of action is illegal or simply wrong. An accounting activity may be legal, but still the wrong thing to do.
- **Ethics** refers to the choices of proper conduct made by an individual in his or her relationships with others. Ethical behavior refers to behavior that is considered "right" or the "right thing to do."
- How individuals determine what constitutes ethical behavior can be influenced by their cultural background, religious views, professional training, and their own moral code.
- **Ethical Guidelines:**
  1. Is it legal?
  2. Does it hurt anyone?
  3. Am I being honest?
  4. Would I care if it happened to me?
  5. Would I publicize my action?
- Ethical behavior is always important to responsible individuals as well as their organizations. There are rules that must be followed if a manager's financial records are to be trusted and if the interpretations made about that financial data is to be perceived as honest.

Key Terms & Concepts Review
Match the key terms with their correct definitions.

1. Hospitality   _____   a. A person skilled in the recording and reporting of financial transactions.
2. Accounting   _____   b. The money paid to a local taxing authority based upon the amount of revenue a hotel achieves when selling its guest rooms.
3. Accountant   _____   c. Those items owned by the business.

4. Financial accounting _____ d. The money that a business must collect from customers and pay to taxing authorities as a result of realizing taxable sales.

5. Revenue _____ e. The cost of the items required to operate the business.

6. Expense _____ f. The money that a business must pay to taxing authorities on individuals employed by the business.

7. Profit _____ g. An individual recognized and certified as highly competent and professional in one or more of the branches of accounting.

8. Assets _____ h. The choices of proper conduct made by an individual in his or her relationships with others.

9. Liabilities _____ i. Individuals who are directly employed by a company to examine that company's own accounting procedures.

10. Owners' equity _____ j. The process of recording financial transactions, summarizing them, and then accurately reporting them.

11. Cost accounting _____ k. The standardized accounting procedures for the restaurant industry.

12. Cost _____ l. The friendly and charitable reception and entertainment of guests or strangers. Also refers to a specific segment of the travel and tourism industry.

13. Tax _____ m. The residual claims owners have on their assets, or the amount left over in a business after subtracting its liabilities from its assets.

14. Tax accounting _____ n. The person responsible for managing a hotel's accounting processes.

15. Occupancy tax _____ o. The system of recording and analyzing transactions for the purpose of making management decisions.

16. Sales tax _____ p. The dollars that remain after all expenses have been paid.

17. Payroll taxes _____ q. The branch of accounting that is concerned with the classification, recording, and reporting of business expenses.

18. Audit _____ r. The standardized accounting procedures for the lodging industry.

19. Auditor _____ s. The standardized accounting procedures for the club industry.

20. Sarbanes-Oxley Act (SOX) _____ t. The individual or group of individuals that completes an independent verification of the financial records of a business.

21. Internal auditors _____ u. Compare and match accounting transactions.

22. External auditors _____
23. Controller _____
24. Reconcile _____
25. Managerial accounting _____
26. Certified Public Accountant (CPA) _____
27. Certified Management Accountant (CMA) _____
28. Hospitality Financial and Technology Professionals (HFTP) _____
29. Hospitality accounting _____
30. Uniform system of accounts _____
31. Uniform System of Accounts for the Lodging Industry (USALI) _____
32. Uniform System of Accounts for Restaurants (USAR) _____
33. Uniform System of Financial Reporting for Clubs (USFRC) _____
34. Ethics _____

v. A charge levied by a governmental unit on income, consumption, wealth, or other basis.

w. Time or resources expended by the business. Also referred to as expense.

x. An independent verification of the financial records of a business.

y. Individuals or firm hired specifically to give an independent (external) assessment of a company's compliance with standardized accounting practices.

z. An organization that offers its own certifications and global network for hospitality professionals working in the accounting and technology areas.

aa. Technically known as the Public Company Accounting Reform and Investor Protection Act, the law provides criminal penalties for those found to have committed accounting fraud.

bb. A series of suggested (uniform) accounting procedures which represent agreed upon methods of recording financial transactions within a specific industry segment.

cc. The term used to indicate the dollars taken in by the business in a defined period of time. Often referred to as sales.

dd. The branch of accounting that is concerned with recording, summarizing, and reporting financial transactions.

ee. An individual recognized and certified as a highly competent professional that assists businesses by integrating accounting information into the business decision process.

ff. The branch of accounting that is concerned with the proper and timely filing of tax payments, forms, or other required documents with the governmental units that assess taxes.

gg. A very specialized area that focuses on those accounting techniques and practices used in restaurants, hotels, clubs, and other hospitality businesses.

hh. The amount of money the business owes to others.

Discussion Questions

1. Describe at least one way in which an owner, an investor, and a manager would use accounting in the hospitality industry.

2. List and briefly explain the five branches of accounting in the hospitality industry.

3. Define a uniform system of accounts and list three systems which are used by the hospitality industry.

4. List five guidelines for ethical behavior by managers and accountants.

Quiz Yourself
Choose the letter of the best answer to the questions listed below.

1. A person who is skilled in the recording and reporting of financial transactions is called a(n):
   a. Manager
   b. Accountant
   c. Cashier
   d. Auditor

2. The formula for calculating profit is:
   a. Revenue - expenses
   b. Revenue + expenses
   c. Revenue / expenses
   d. Revenue x expenses

3. The balance sheet equation is expressed as:
   a. Assets = Liabilities – Owner's Equity
   b. Assets = Liabilities / Owner's Equity
   c. Assets = Owner's Equity – Liabilities
   d. Assets = Liabilities + Owner's Equity

4. Business accountants who are skilled at recording, summarizing, and reporting financial transactions specialize in:
   a. Managerial accounting
   b. Auditing
   c. Financial accounting
   d. Cost accounting

5. Business accountants who are concerned with the classification, recording and reporting of business expenses specialize in:
   a. Managerial accounting
   b. Auditing
   c. Financial accounting
   d. Cost accounting

6. The branch of accounting which is chiefly concerned with the accuracy and truthfulness of financial reports is called:
   a. Managerial accounting
   b. Auditing
   c. Financial accounting
   d. Cost accounting

7. The Sarbanes-Oxley Act (SOX) was passed in 2002 and provides for:
   a. Tax credits for companies that are financially unsound
   b. Criminal penalties for those found to have committed accounting fraud
   c. Tax penalties for companies that are financially sound
   d. Criminal penalties for those found to be practicing accounting without a license

8. A uniform system of accounts developed for restaurants is called the:
   a. USAR
   b. HFTP
   c. USALI
   d. SOX

9. A uniform system of accounts developed for the lodging industry is called the:
   a. USAR
   b. HFTP
   c. USALI
   d. SOX

10. Ethics refers to:
    a. Agreed upon methods of reporting financial transactions
    b. Specialized knowledge regarding hospitality financial activities
    c. Actions which may be wrong, but are nevertheless legal
    d. Choices of proper conduct made by an individual in his or her relationships with others.

Chapter 2

Accounting Fundamentals Review

Highlights

* Bookkeeping and Accounting
* The Accounting Formula
* Recording Changes to the Accounting Formula
* Generally Accepted Accounting Principles
* The Hospitality Business Cycle
* Key Terms and Concepts Review
* Discussion Questions
* Quiz Yourself

Study Notes

Bookkeeping and Accounting

- In the hospitality industry, **bookkeepers** of all types perform the critically important task of initially recording financial transactions in a business.
- If the bookkeeping tasks of the servers, bartenders, kitchen staff, and managers of a restaurant, and the front desk, controller and other staff of a hotel are not properly performed, the resulting financial data generated by these business's accountants will not be accurate and decisions made based upon the numbers supplied are likely to be flawed.
- Bookkeeping forms the foundation of accurate financial reporting and analysis (see Figure 2.2).
- As a hospitality manager it is important that you ensure accurate and timely bookkeeping and accounting methods to produce the financial data you must analyze to make decisions.

The Accounting Formula

- An **account** is a device used to record increases or decreases in the assets, liabilities or owners' equity portion of a business.
- There is a very specific and unchanging relationship between assets, liabilities, and owners' equity. This relationship is expressed in a mathematical formula so precise, clear-cut, and unchanging that it is actually called "**The Accounting Formula**". The Accounting Formula states that, for every business:

$$\text{Assets} = \text{Liabilities} + \text{Owners' Equity}$$

- Using basic algebra, variations of this formula can be developed:

> **Assets = Liabilities + Owners' Equity**
>
> **Assets - Liabilities = Owners' Equity**
>
> **Assets - Owners' Equity = Liabilities**

- Owners' equity accounts include two major sub-categories called permanent accounts and temporary accounts. **Permanent owners' equity accounts** include items such as stock (or owner's investment) and **retained earnings** (accumulated account of profits over the life of the business that have not been distributed as dividends).
- **Dividends** are money paid out of net income to stockholders as a return on their investment in the company's stocks.
- **Temporary owners' equity accounts** include revenue (increase owners' equity) and expense accounts (decrease owners' equity). At the end of the accounting period, the temporary accounts are **closed out** (their balances reduced to zero). The resulting current period's net profit or loss is used to update the balance of the permanent owners' equity account (in retained earnings).
- The permanent and temporary owners' equity accounts are shown in the following modification of The Accounting Formula:

> **Assets =   Liabilities**
> **+ Permanent Owners' equity (Stocks + Retained Earnings)**
> **+ Temporary Owners' equity (Revenue - Expenses)**

- The balance sheet and the income statement are developed from The Accounting Formula. The **balance sheet** is an accounting summary that closely examines the financial condition of a business, by reporting the value of a company's total assets, liabilities and owners' equity on a specified date.
- The **income statement** reports in detail and for a very specific time period, a business's revenue from all its revenue producing sources, the expenses required to generate those revenues, and the resulting profits or losses (net income).

Recording Changes to the Accounting Formula

- Every time a business makes a financial transaction, it has an effect on (changes) The Accounting Formula. Accountants must report changes to the Accounting Formula while always ensuring that the formula stays in balance. (See Figure 2.3)
- Additions to or subtractions from one of the sides of the scale *must* be counterbalanced with an equal addition to, or subtraction from, the other side of the scale if the formula is to stay in balance, and the scale is to remain in its mandatory "equal" position.

- It is also possible to make changes (additions or subtractions) to only one side of the scale. For example, an equal dollar value added to and then subtracted from the asset total would not cause the overall formula to be out of balance.

Double-Entry Accounting

- **Double entry accounting** (sometimes called double entry bookkeeping) requires that the person recording a financial transaction make at least two separate accounting entries (changes to its accounts) every time a financial transaction modifies The Accounting Formula of a business.
- A double entry system is used to catch recording errors and to accurately track the various streams of money in and out of businesses.

The Journal and General Ledger

- When utilizing the double entry accounting system, each of a business's individual transactions are originally recorded in the business's unique journal of financial transactions. A **journal**, then, is the written record of a specific business's financial transactions. A **journal entry** is made to a specific account when changes to The Accounting Formula are recorded.
- The up-to-date balances of all a business's individual asset, liability, and owners' equity, (as well as revenue and expense) accounts are maintained in its **general ledger**.
- Important concepts for an accountant to remember about maintaining a business's general ledger are:
  - The Accounting Formula, which is the summary of a business's asset, liability and owners' equity accounts, must stay in balance.
  - The Accounting Formula is affected every time a business makes a financial transaction.
  - Each financial transaction is to be recorded two times in a double entry accounting system.
  - The original records of a business's financial transactions are maintained in its journal, and each financial transaction recorded is called a journal entry.
  - The current balances of each of a business's individual asset, liability and owners' equity accounts are totaled and maintained in its general ledger.

Credits and Debits

- The asset, liability and owners' equity portions of the Accounting Formula can be broken down into smaller units called accounts. Because of their shape, accountants often call these individual accounts **"T" accounts**:

```
          Name of Account
_____
                    |
 Left (Debit)       |   Right (Credit)
                    |
```

- A T account consists of three main parts:
  1. The top of the T is used for identifying the name of the account.
  2. The left side of a T account is called the **debit** side. Each journal entry made on the *left* side of a T account is always called a **debit entry**.
  3. The right side of a T account is called the **credit** side. Each journal entry made on the *right* side of a T account is always called a **credit entry**.
- The manner in which an accountant uses T accounts to make a journal entry can be conceptualized in much the same way as the scale in The Accounting Formula.
- Within each of the three major components of The Accounting Formula, accountants create individual T accounts to clarify the financial standing of the business. Some of these accounts include:
  - **Depreciation** is a method of allocating the cost of a fixed asset over the useful life of the asset. Once fully depreciated, the value of the asset at the end of its useful life is called its **salvage value**.
  - **Accumulated depreciation** is a record and accumulation of all depreciation expense charges that occur over the life of the asset. It is listed as a **contra asset** and represents deductions to a fixed asset. Contra assets, like accumulated depreciation, behave *opposite* of all other asset accounts with regard to debits and credits.
  - **Accounts receivable** (often shortened to AR) represent the amount of money owed to a business *by others* (such as customers) and thus is considered to be one of that business's asset accounts.
  - **Accounts payable** (often shortened to AP) represents the amount of money owed by the business *to others* (such as suppliers), and as a result is considered to be one of that business's liability accounts.
- The difference between a T account's total debits and total credits is called the **account balance**. A T account set up to monitor the value of an asset account will, in most cases, have a debit balance. This is so because additions to the current balance of an asset account are (simply by tradition) recorded on the left (debit) side of a T account. Reductions in the value of an asset account are recorded (again simply by tradition) on the right (credit) side of its T account.
- Debit and credit entries have a different impact on each of the three major components of The Accounting Formula (see Figure 2.9).

- Remembering the impact of debits and credits on the components of The Accounting Formula can be difficult. Don St. Hilaire and his students at California State Polytechnic University, Pomona devised a "trick" that will help make this easier. Imagine that your left hand represents debits and your right hand represents credits.
- Your fingers represent the following accounts, where the left hand shows debits and the right hand shows credits:

    Thumb -   Assets (A)

    Pointer -   Liabilities (L)

    Middle finger -   Owners' Equity (OE)

    Ring finger -   Revenues (R)

    Pinky finger -   Expenses (E)

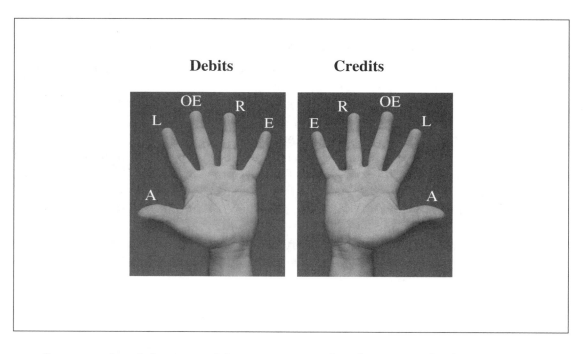

- Increases ↑ and decreases ↓ in accounts can then be summarized:
    o **Debits** = ↑ Assets, ↓ Liabilities, ↓ Owners' Equity, ↓ Revenues, ↑ Expenses
    o **Credits** = ↓ Assets, ↑ Liabilities, ↑ Owners' Equity, ↑ Revenues, ↓ Expenses
- Using the classification of "fingers" and the increases and decreases in accounts shown above, your hands can be used to remember the impact of debit and credit entries on The Accounting Formula components. Fingers "up" represent increases and fingers "down" represent decreases:

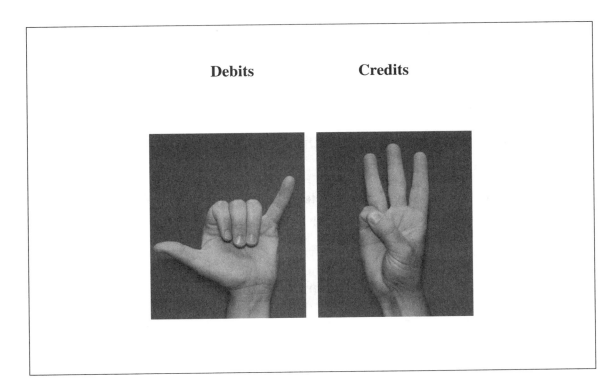

- Two of the most important and frequently used T accounts are revenue and expense accounts. These two account types belong to the owners' equity portion of the accounting equation and are summarized, and closed out, at the end of each accounting period.

Generally Accepted Accounting Principles

- Professionals in the field of accounting have worked hard to develop and consistently follow **generally accepted accounting principles (GAAP)**, in order to describe the best method of recording any financial transaction and to ensure that readers of financial statements can immediately depend upon their accuracy.
- Generally accepted accounting principles are developed by the **Financial Accounting Standards Board (FASB), a** private body whose mission is to establish and improve standards of financial accounting and reporting for the guidance and education of the public, including issuers, auditors and users of financial information. As a result of their importance, accountants, auditors and controllers all utilize GAAP.
- Eleven of the most critical generally accepted accounting principles all hospitality managers simply must recognize include:
  - Distinct business entity principle
  - Going concern principle
  - Monetary unit principle
  - Time period principle
  - Cost principle
  - Consistency principle

- Matching principle
- Materiality principle
- Objectivity principle
- Conservatism principle
- Full disclosure principle

The Distinct Business Entity Principle

- The **distinct business entity principle** states that a business's financial transactions should be kept completely separate from those of its owners.
- There are three basic types of business ownership in the United States:
  - A **corporation**, commonly called a **C corporation**, is a legal entity that is separate and distinct from its owners. It is allowed to own assets, incur liabilities and sell shares of ownership, among other things.
    - A **Limited Liability Corporation (L.L.C.)** is a special form of a corporation that is typically regulated by the state in which it is formed. An LLC limits the potential losses incurred by its owners only to what they have invested in the business.
    - A **Sub S corporation** is another type of corporation that is granted special status under U.S. tax laws. These laws are very specific about how and when this type corporation can be formed and the number of **stockholders** (company owners) it can have.
  - A **partnership** is simply a business entity where two or more individuals agree to share ownership. Its profits are taxed differently than those profits earned by a corporation.
    - In a **Limited Partnership (LP)** one or more general partners manage the business and are liable for its debts and one or more limited partners invest in the business but have limited personal liability for its debts.
    - A **proprietorship** is a business owned by a single individual. A sole proprietor pays personal (but not corporate) income tax on profits made by the business and also has unlimited liability for the debts and other obligations incurred by the business.

The Going Concern Principle

- The **going concern principle** means that accountants make the assumption that the business will be ongoing (continue to exist) indefinitely and that there is no intention to **liquidate** (sell) all of the assets of the business.
- The going concern principle clearly directs accountants to record the value of a business's assets only at the price paid for them, so that readers of a financial statement know that asset values represent a business's true cost, and not the cost of liquidation or replacement.

The Monetary Unit Principle

- The **monetary unit principle** means that financial statements must be prepared in a specific currency denomination. In the United States, the U.S. dollar is the monetary unit used for preparing financial statements.
- Fulfilling the monetary unit principle can be quite complicated, since companies often operate in more than one country, and use more than one currency in their operating transactions.

The Time Period Principle

- The **time period principle** requires a business to identify the time period for which its financial transactions are reported.
- A **fiscal year** consists of 12 consecutive months (but not necessarily beginning in January and ending in December like a **calendar year**).
- A fiscal year-end financial report would provide owners with the information they need to file their taxes and make other needed financial management decisions.
- The amount of time included in any summary of financial information is called an **accounting period**. The managers of a business may be most interested in monthly, weekly, or even daily financial summary reports.

The Cost Principle

- The **cost principle** requires accountants to record all business transactions at their cash cost.
- Just as the going concern principle requires accountants to value a business's assets at their purchase price, with few exceptions, it requires businesses to set the value of the items it intends to sell at the price the business actually paid for them.

The Consistency Principle

- The **consistency principle** of accounting states that a business must select and consistently report financial information under the rules of the specific reporting system it elects to use.
- In an **accrual accounting** system, revenue is recorded when it is earned, regardless of when it is collected, and expenses are recorded when they are incurred, regardless of when they are paid. The result is a more accurate reflection of a business's true *monthly* profitability.
- A **cash accounting** system records revenue as being earned when it is actually received and records expenditures when they are actually paid, regardless of when they were incurred.
- Most hotels and restaurants choose to use the accrual accounting system, recording their revenue when their customers make purchases, rather than when these customers actually pay their bills.

- Similarly, when an expense such as a tax payment is paid at one time for an entire year, rather than recording it when it is actually paid, in the accrual accounting system, 1/12 of the tax expense is assigned to each of the 12 monthly revenue and expense summaries prepared.

The Matching Principle

- The **matching principle** is designed to closely match expenses incurred to the actual revenue those expenses helped generate. This principle applies to those organizations that elect to use an accrual system of accounting.

The Materiality Principle

- The consistency and matching principles require accountants to expense the cost of certain long-life assets like furniture and equipment over the time period in which they will help a business generate revenue. The **materiality principle**, however, allows accountants, under very strict circumstances, to vary from these two important principles.
- The **materiality principle** means that if the value of an item is deemed to be not significant, then other accounting principles may be ignored if it is not practical to use them.

The Objectivity Principle

- The **objectivity principle** states that financial transactions must have a confirmable (objective) basis in fact. While all of the generally accepted accounting principles have great value, it is this one that most ensures that a business's financial statements can be trusted to be reliable.
- Sales should have substantiating evidence to prove that they actually occurred, such as guest checks, bank card statements, or various sales records maintained in an electronic cash register or computer.
- Before they can be recorded as having been incurred or paid, expenses must be verified with evidence such as delivery slips or original invoices supplied by vendors, cancelled checks or documented **electronic funds transfers (EFTs).**

The Conservatism Principle

- The **conservatism principle** requires the accountants of a business to be conservative when reporting its revenue (and thus not to report it until it is actually earned) and realistic when reporting its expense and other liabilities.

The Full Disclosure Principle

- The **full disclosure principle** requires that any past or even *future* event which could materially affect the financial standing of the business and that cannot be easily discerned from reading the business's financial statements must be separately reported.
- These reports, prepared in the form of footnotes, must be attached to the financial statements prepared by the business's accountants.
- Accountants use full disclosure footnotes to report events that have not happened yet but may, if they do occur, considerably change the conclusions drawn by readers of a business's financial statements. Examples of such events include significant lawsuits, changing from a cash to an accrual accounting system, significant tax disputes, modifying depreciation schedules, unusual events that occurred after the financial statements were actually prepared, or any other atypical or non-recurring event that could materially affect the business.

The Hospitality Business Cycle

- An accounting period is the time frame included in the financial transaction summaries prepared by a business. Regardless of the time frame involved, each of these accounting summaries will consist of common components that make up a complete **hospitality business cycle** (see Figure 2.14).
- Business cycles most often begin by utilizing cash to secure raw materials for processing by workers who create finished products (meals are produced, and guest rooms are prepared). The sale of these products creates cash or accounts receivables owned by the business. The money resulting from these sales is then used to buy additional products or, if profits are generated, some of it may be retained by the business. Then the business cycle begins again.

Key Terms & Concepts Review
Match the key terms with their correct definitions.

1. Bookkeepers _____ a. Assumption that the business will be ongoing (continue to exist) indefinitely and that there is no intention to liquidate all of the assets of the business.

2. Account _____ b. Twelve consecutive months beginning in January and ending in December.

3. The Accounting Formula _____ c. These accounts include revenue and expense accounts which can increase owners' equity (revenue accounts) or decrease owners' equity (expense accounts).

4. Permanent owners' equity accounts _____
5. Retained earnings _____
6. Dividends _____
7. Temporary owners' equity accounts _____
8. Close out _____
9. Balance sheet _____
10. Income statement _____
11. Double entry accounting _____
12. Journal _____
13. Journal entry _____
14. General ledger _____
15. T accounts _____
16. Debit _____
17. Debit entry _____
18. Credit _____
19. Credit entry _____
20. Depreciation _____

d. Money paid out of net income to stockholders as a return on their investment in the company's stocks.
e. A business entity where two or more individuals agree to share ownership.
f. Managers or owners of a hotel.
g. To be understandable, financial statements must be prepared in an identifiable monetary unit (specific currency denomination).
h. A report that details for a very specific time period, a business's revenue from all its revenue producing sources, the expenses required to generate those revenues, and the resulting profits or losses (net income).
i. The process of reducing temporary accounts (revenue and expense) to zero.
j. Requires accountants to record all business transactions at their cash cost.
k. Money that is paid or moved electronically from the business to the entity to whom money is owed.
l. A journal entry made on the right side of a T account.
m. The differences between a T account's total debits and total credits.
n. A method of allocating the cost of a fixed asset over the useful life of the asset.
o. The amount of money owed by the business to others, such as suppliers.
p. Accumulated account of profits over the life of the business that have not been distributed as dividends.
q. Individual accounts shaped like a "T" used in a double entry accounting system.
r. A record of increases or decreases in the assets, liabilities, or owners' equity of a business.
s. A journal entry made on the left side of a T account.
t. A legal entity that is separate and distinct from its owners and is allowed to own assets, can incur its own liabilities, and can sell shares of ownership.

21. Salvage value _____
22. Accumulated depreciation _____
23. Contra asset _____
24. Accounts receivable _____
25. Accounts payable _____
26. Account balance _____
27. Folios _____
28. Generally Accepted Accounting Principles (GAAP) _____
29. Financial Accounting Standards Board (FASB) _____
30. Distinct business entity principle _____

u. The distinctive type of corporation that is granted special status under United States tax laws which are very specific about how and when this type corporation can be formed and the number of stockholders it can have.

v. Persons who perform the task of initially recording financial transactions of a business.

w. Managers or owners of a restaurant.

x. A business entity in which one or more general partners manage the business and are liable for its debts and one or more limited partners invest in the business but have limited personal liability for its debts.

y. Twelve consecutive months not necessarily beginning in January and ending in December like a calendar year.

z. Procedure requiring that a financial transaction has at least two separate accounting entries (changes to its accounts) every time a financial transaction modifies The Accounting Formula of a business.

aa. A system in which revenue is recorded when it is earned, regardless of when it is collected, and expenses are recorded when they are incurred, regardless of when they are paid.

bb. A private body whose mission is to establish and improve standards of financial accounting and reporting for the guidance and education of the public, including issuers, auditors, and users of financial information. It publishes its recommendations known as generally accepted accounting principles.

cc. A special form of a corporation that limits the potential losses incurred by its owners only to what they have invested in the business and is typically regulated by the state in which the company is formed.

dd. An accounting summary that closely examines the financial condition, or health, of a business by reporting the value of a company's total assets, liabilities, and owners' equity on a specified date.

31. Corporation _____   ee. A system that records revenue as being earned
    (C corporation)   when it is actually received and expenditures
                               when they are actually paid, regardless of when
                               they were incurred.

32. Limited Liability _____ ff. If an item is deemed to be not significant
    Corporation (LLC)   (material), then other accounting principles may
                                 be ignored if it is not practical to use them.

33. Sub S corporation _____ gg. A business must select and consistently report
                                 financial information under the rules of the
                                 specific system it elects to use.

34. Stockholders _____ hh. Individual bills for a hotel guest or room.

35. Partnership _____ ii. The written record of a specific business's
                             financial transactions.

36. Limited Partnership _____ jj. Requires that any past or even future event
    (LP)   which could materially affect the financial
                                    standing of the business and that cannot be
                                    easily discerned from reading the business's
                                    financial statements must be separately reported
                                    usually in the form of footnotes.

37. Proprietorship _____ kk. The amount of time included in any summary
                                of financial information.

38. Going concern _____ ll. Requires the accountants of a business to be
    principle   conservative when reporting its revenue (and
                              thus not to report it until it is actually earned)
                              and realistic when reporting its expenses and
                              other liabilities.

39. Liquidate _____ mm. A record of the up-to-date balances of all a
                          business's individual asset, liability, and
                          owners' equity, (as well as revenue and
                          expense) accounts.

40. Monetary unit _____ nn. Represents deductions to a fixed asset and
    principle   behaves *opposite* of all other asset accounts with
                              regard to debits and credits.

41. Time period _____ oo. Standards used to develop financial statements
    principle   that provide consistency and accuracy in
                            reporting financial information.

42. Fiscal year _____ pp. Closely matches expenses incurred to the
                            actual revenue those expenses helped generate.

43. Calendar year _____ qq. Owners which hold shares of stocks in a
                              company.

44. Accounting period _____ rr. The amount of money owed to a business by
                                  others, such as customers.

45. Cost principle _____ ss. Assets = Liabilities + Owners' Equity.

46. Accrual accounting _____ tt. Requires a business to clearly identify the time period for which its financial transactions are reported.

47. Cash accounting _____ uu. Financial transactions must have a confirmable (objective) basis in fact, and there must be a way to verify that a financial transaction actually occurred before it can be recorded in the business's financial records.

48. Consistency principle _____ vv. The left side of a T account.

49. Matching principle _____ ww. The principle states that a business's financial transactions should be kept completely separate from those of its owners.

50. Materiality principle _____ xx. The estimated value of an asset at the end of its useful life.

51. Objectivity principle _____ yy. The right side of a T account.

52. Electronic funds transfer (EFT) _____ zz. These accounts include stock (or owner's investment) and retained earnings.

53. Conservatism principle _____ aaa. A record and accumulation of all depreciation expense charges that occur over the life of an asset.

54. Full disclosure principle _____ bbb. Record made to a specific account when changes to The Accounting Formula are documented.

55. Restaurateurs _____ ccc. Sell assets of a business.

56. Hoteliers _____ ddd. A business owned by a single individual.

Discussion Questions

1. List five concepts to remember about maintaining a business's general ledger.

   _____
   _____
   _____

2. Describe the three main parts of a "T" account.

   _____
   _____
   _____

3. Describe the hand "trick" and list the accounts represented by each finger.
_____
_____
_____

4. List five of the eleven generally accepted accounting principles.
_____
_____
_____

Quiz Yourself
Choose the letter of the best answer to the questions listed below.

1. The Accounting Formula states that, for <u>every</u> business
   a. Assets + Owner's Equity = Liabilities
   b. Assets = Owner's Equity - Liabilities
   c. Assets + Liabilities = Owner's Equity
   d. Assets = Liabilities + Owner's Equity

2. Permanent owner's equity accounts include
   a. Stock and retained earnings
   b. Retained earnings and dividends
   c. Stock and dividends
   d. Stock and assets

3. An accounting summary that reports the value of a business's total assets, liabilities and owner's equity on a specific date is the
   a. Income Statement
   b. Dividend Statement
   c. Cash Flows Statement
   d. Balance Sheet

4. The accounting system that requires at least two separate accounting entries for every financial transaction is called
   a. Two entry accounting
   b. Double entry accounting
   c. General ledger accounting
   d. The Accounting Formula

5. The up-to-date balances of all of a business's asset, liability, owner's equity, revenue and expense accounts are maintained in the
   a. Company journal
   b. Balance sheet
   c. General ledger
   d. Double entry ledger

6. On a T account, the following is true
   a. Each journal entry made to the right side of a T account is called a credit entry
   b. Each journal entry made to the left side of a T account is called a credit entry
   c. Each journal entry for an asset account is made to the right side of its T account
   d. Each journal entry for a revenue account is made to the right side of its T account

7. The difference between a T account's total debits and total credits is called the
   a. Credit balance
   b. Account balance
   c. money balance
   d. Debit balance

8. What type of account entry can be summarized as having the following impact on the components of The Accounting Formula: ↑ Assets, ↓ Liabilities, ↓ Owners' Equity, ↓ Revenues, ↑ Expenses
   a. Increases
   b. Decreases
   c. Credits
   d. Debits

9. The best method of recording any financial transaction and ensuring that readers of financial statements can depend on their accuracy is to follow
   a. The T account system
   b. The double entry accounting system
   c. Generally accepted accounting principles
   d. The direct method of accounting

10. When steaks are purchased with cash, cooked and then sold to the public, and the resulting revenues are then used (among other things) to purchase more steaks, this is an example of
    a. Hospitality business cycle
    b. T account system
    c. Balance sheet
    d. Double entry accounting system

Part II: Financial Statements

Chapter 3

The Income Statement

Highlights

- * The Purpose of the Income Statement
- * Income Statement Preparation
- * Income Statement Analysis
- * Key Terms and Concepts Review
- * Discussion Questions
- * Quiz Yourself

Study Notes

The Purpose of the Income Statement

- When you manage a hospitality facility, you will receive **revenue**, the term used to indicate the money you take in, and you will incur **expenses**, the cost of the items required to operate the business. The dollars that remain after all expenses have been paid represent your **profit.**

> **Revenue – Expenses = Profit**

- Many hospitality managers call each individual revenue generating segment within their business a **profit center**.
- The revenue-expense = profit formula holds even in what is not typically considered a for-profit segment of the hospitality industry.
- In many **business dining** situations, food is provided as a service to the company's employees either as a no-cost (to the employee) benefit or at a greatly reduced price.
- Thus, it is common in many situations to operate a **cost center** that generates costs but no revenue.
- Managers are not the only ones interested in a business's revenue, expenses and profits. All **stakeholders** who are affected by a business's profitability will care greatly about the effective operation of a hospitality business. These stakeholders may include:
  - o Owners
  - o Investors

- o Lenders
- o Creditors
- o Managers
- When an accurate income statement is used to provide information, the business's owners, lenders, investors and managers can all make better decisions about how best to develop and operate it.

Owners

- The owners of a business typically have the greatest interest in its success. In many companies, however, the owners do not actively participate in the management of the business.
- By evaluating the data in the income statement, those responsible for managing a business will better be able to determine the effectiveness of the manager they have selected and the progress made toward goal achievement.

Investors

- Investors supply funds to restaurants and hotels to earn money on their investment. These earnings generally include periodic cash payments from profits generated by the business, plus any property **appreciation** (increase in value) achieved during the period of the investment.
- Restaurants, hotels, and many other hospitality businesses are inherently risky investments because they require specialized management expertise, can be significantly subject to economic upturns and downturns, and are often not easy to sell rapidly.
- Before individuals, corporations, or other financial entities elect to invest in a business, they will want to know that their investment is a good one.
- One way to evaluate the quality or strength of an investment is to measure the return on investment it generates. **Return on investment (ROI)** is simply a ratio of the money made compared to the money invested. ROI is computed as follows:

$$\frac{\text{Money earned on funds invested}}{\text{Funds invested}} = \text{ROI}$$

- This ratio of earnings achieved to investment is critical to determining investment quality. For an example, see *Go Figure!* in the textbook.
- As a general rule, the higher the ROI sought by an investor, the more risky is the investment. Different investors, with different investment goals, will evaluate and choose from a variety of investment options.
- Computing the amount of money that is (or may be) earned on invested funds requires a careful reading of a business's income statement, because the income statement is the source of the information required to compute ROI.

Lenders

- Most hotel or restaurant acquisitions are financed with both **debt** (funds lent to a business) and **equity** (funds supplied by its investors or owners).
- Lenders to businesses in the hospitality industry include banks, insurance companies, pension funds and other similar capital sources.
- In actuality, there are essentially two types of entities that lend money to hospitality businesses. The first of these are lenders who simply agree to fund or finance a business. Typically, this is done by granting the business a loan that must be repaid under the specific **terms** or conditions of the loan.
- Lenders have first claim to the profits generated by the business. They will be repaid before a business is permitted to distribute profits to its investors.
- The terms of the loan might include the fact that the loan must be repaid, along with 8% **interest**. This interest rate would represent the ROI sought by the bank's managers for their bank's money.
- Like investors, lenders seek greater returns (higher interest rates) when they perceive the risks of a business not repaying its loan are high. Alternatively, when a lender believes the risk of **default** (non-payment) of a loan is low, interest rates charged for the use of the lender's funds will be lower also.

Creditors

- A **creditor** is a company to whom a business owes money, such as a vendor. For example, when a food and beverage director purchases food from a vendor, that vendor is likely to extend the restaurant credit.
- The food ordered by the food and beverage director would be delivered and an **invoice** detailing the purchases, as well as the cost of the food items purchased, would also be delivered.
- The expectation of the vendor, of course, is that the restaurant would pay the vendor for the delivered food according to the credit terms established by the vendor.
- A business's vendors must also be profitable. Vendors are very interested in the creditworthiness of their customers. A company's **creditworthiness**, or the ability to pay bills promptly, is a result of its profitability, which is measured, in large part, by the information contained in the income statement.

Managers

- Many hospitality managers consider the income statement to be a reflection of their managerial ability, for a variety of reasons.
- The income statement details how profitable an operation has been within a designated time period. Excellent managers tend to operate more profitable facilities than do poorer managers.

- In some cases, operational performance as measured by the income statement's results is used to establish managers' raises, compute their bonuses, and in many companies, even determine promotional opportunities.
- Employees are also important stakeholders in a business. If a business is profitable, employees' jobs are more secure than if the business is struggling. Employee wages are more likely to rise, expansion of the business may create additional job opportunities, and employers may have the ability to provide employees with benefit programs that satisfy their needs as well as those of the business.

Income Statement Preparation

- In very small hospitality operations, the owner or managers of the business may be responsible for the preparation of the income statement.
- In larger restaurant operations, and especially those in the quick service restaurant (QSR) industry, the manager may simply submit financial data about a store to a centralized accounting office, which would then prepare the unit's income statement.
- In very large restaurants and in many hotels, the income statement may be prepared by professionals who work on-site.
- In all cases, however, as a manager working in the hospitality industry, you must understand the income statement format utilized, as well as how revenue, expenses and profits are reported.

Format

- An income statement is designed to identify revenues, expenses, and profits. An income statement is a summary of financial information for a defined accounting period (see Figure 3.1 for an example).
- When an operation's revenue exceeds its expenses, the amount of the profits (revenue – expense) is simply presented on the statement.
- Negative numbers (losses) on an income statement are designated one of three ways:
  o By a minus "-" sign in front of the number.
  o By brackets "( )" around the number.
  o By the use of the color red (rather than black) to designate the loss amount. This approach gives rise to the slang phrase to "**operate in the red**" to indicate a business that is not making a profit. Similarly, to "**operate in the black**" would indicate the business is profitable.
- In the hospitality industry the physical layout of the actual income statements for different types of business will be somewhat different.
  o For example, a restaurant that serves alcohol would, on its income statement, identify the revenue generation and costs associated with serving drinks. In a restaurant that serves only food and non-alcoholic beverages, the income statement would not include a section for alcoholic beverages.

- At a limited service hotel that generates its revenue only from room sales, telephone toll calls, and the sale of in-room movies, the income statement would reflect those revenue (profit) centers. In a full-service resort hotel, multiple food and beverage outlets and diverse retail profit centers would be included on the income statement.
- A restaurant income statement, using the Uniform System of Accounts for Restaurants (USAR), shows sales and cost of sales related to food and beverage and any other expenses related to the functioning of the restaurant.
- The USAR can be divided into three sections: gross profit, operating expenses, and nonoperating expenses, arranged on the income statement *from most controllable to least controllable* by the foodservice manager.
  - The **gross profit section** consists of food and beverage sales and costs that can and should be controlled by the manager on a daily basis.
  - The **operating expenses section** is also under the control of the manager but more so on a weekly or monthly basis (with the exception of wages, which you can control daily).
  - The third section of the USAR is the **nonoperating expenses section.** It is this section that is least controllable by the foodservice manager.
- Figures 3.3 and 3.4 show two alternative formats for a hotel income statement using the Uniform System of Accounts for the Lodging Industry (USALI).
- The USALI can be divided into three sections: operated department income, undistributed operating expenses, and nonoperating expenses.
  - The **operated department income section** consists of separate profit centers as department income (rooms, food, beverage, telecommunications, other departments, and rental and other income). Each department reports revenues, expenses, and income.
  - The **undistributed operating expenses section** covers Undistributed Operating Expenses through Gross Operating Profit. This section includes expenses that cannot truly be assigned to one specific department, such as security, transportation, and franchise fees.
  - The **nonoperating expenses section** includes Rent, Property Taxes, and Insurance through Net Income. This section includes depreciation, interest, and income taxes, and is least controllable by the hotel manager.
- Depreciation expense (see Chapter 2) in all forms of the income statement serves a very specific purpose. Depreciation is subtracted from the income statement primarily to lower income, and thus lower taxes.
- While the layout of a business's income statement will vary somewhat based upon the Uniform System of Accounts applicable to its industry segment, there are basic commonalities among income. The process of preparing an income statement for a business consists of the three following steps:
  1. Identification of accounting period
  2. Documentation of revenue data
  3. Documentation of expense data

Accounting Period

- The period of time (the specific number of days and dates) included in a summary of financial information is called an accounting period.
- For many businesses, the accounting periods established coincide with the calendar months of the year.
- In some cases, businesses prefer to create income statements that are 28 days long, because these managers seek to create perfectly "equal" accounting periods. This helps the manager compare performance from one period to the next without having to compensate for "extra days" in any one period.
- Businesses may also elect to create income statements bi-monthly, quarterly or even annually (see Figure 3.5).
- In some cases, managers who utilize longer accounting periods may also be interested in "mini" income statements that detail specially selected revenue and expense categories on a shorter term basis (weekly, daily, or even hourly
- The income statement time must be clearly indicated at the top of the prepared income statement document, to ensure that the document's readers know, *before* they begin reading, exactly when the financial period which is being reported upon actually began and ended.

Revenue Data

- The first portion of the income statement details the revenue data to be reported during the identified accounting period.
- In many businesses, revenue listed on the income statement will come from more than one source. In a large resort hotel, there might be dozens of sources of revenue. They are frequently listed using the procedures recommended by the Uniform System of Accounts for the Lodging industry (USALI).
- Currently, the USALI suggests that managerial accountants in the hotel industry use one or more of the following categories to record their hotel's revenues:
    - Rooms
    - Food
    - Beverage
    - Telecommunications
    - Garage and Parking
    - Golf Course
    - Golf Pro Shop
    - Guest Laundry
    - Health Center
    - Swimming Pool
    - Tennis
    - Tennis Pro Shop
    - Other Operated Departments
    - Rentals and Other Income

- Hotels with special revenue generating features may add additional revenue reporting sections to record that special feature revenue on their income statements.
- On the other extreme, a hospitality operation may simply have one source of revenues.

Expense Data

- Following the title, accounting period and revenue of the income statement is the section reserved for identifying the expenses incurred by the business.
- Two major issues must be well-understood when considering the expense data included on income statements. These issues are the **timing** and the **classification** or placement of the expense.

Expense Timing

- Accrual accounting requires a business's revenue to be reported when earned and its expenses to be recorded when incurred. This matching principle is designed to closely tie expenses of a business to the actual revenues those expenses helped the business generate.
- Decisions made about how best to match revenues to expenses can be complex and even open to honest difference of opinion among hospitality managers. These decisions will be established by generally accepted accounting principles, by company policy, by property-specific policy, or in some cases, the best decision that can be made by the person(s) preparing the income statement.
- The consistency principle of accounting requires managers to be uniform in decision making. That is, if an expense is treated in a specific manner in one instance, it should be treated in an identical manner in all subsequent situations.

Expense Classification

- **Expense classification** is the process of carefully considering how a business's expenses will be detailed for reporting purposes.
- Managers may seek to place (assign) to each department or profit center the expense that center utilized to generate its revenue. These managers concern themselves not merely with the proper timing of expenses; they are equally concerned about the placement of the expense as well.
- In a small operation, costs may be easy to classify. In a larger operation with multiple profit centers, operating expenses may be harder to classify.
- In the hotel industry, when an expense is easily attributable to one department, it is classified as a **departmental cost.** This type of cost is sometimes referred to as a **direct operating expense**.
- When the expense cannot truly be assigned to one specific area within an operation, it is classified as an **undistributed operating expense**.

Schedules

- In addition to income statement summaries, managerial accountants may use one or more departmental **schedules** to provide statement readers with more in-depth information about important areas of revenues and expenses. The revenue or expense portion of an income statement could consist of only one line or several sources, or it may consist of a summary with reference to one or more departmental schedules that will provide additional detail.
- Uniform System of Accounts exist for restaurants (USAR), the lodging industry (USALI), and clubs (USFRC). The methods managerial accountants in those industries use to create appropriate schedules are important areas addressed by these uniform systems.
- Managers should always strive to use the specific methods of reporting that best maximize and clarify the information provided to the statement's readers.
- Hotels of various types will provide readers of their financial statements the most information if they create schedules geared for their specific operation. Many rooms managers in the industry frequently use one or more of the following segments to classify guests and thus the revenues they generate:
  - Transient (Individual)
    - Corporate
    - Leisure
    - Discount
    - Package
    - Long-term stay
  - Group
    - Corporate
    - Leisure
    - Discount
    - Package
    - Association
    - Tour Bus
    - Airline crew
    - Social, Military, Education, Religious, Fraternal (**SMERF**)
    - Other
- In the food and beverage revenue area, revenue schedules may be created based upon the sales achieved by individual restaurants and bars within the hotel, guest types, or even hours of operation. The purpose of each revenue or expense schedule created should be that of enhancing and clarifying the reader's understanding of the income statement.
- It is more critical to understand the principles behind the preparation of an income statement than the current format of any specific business's income statement (see Figure 3.9).

Income Statement Analysis

- Income statements summarize data. Managers analyze that data because they rely on their income statements to help them better understand, and thus better manage, their business. A manager will be primarily interested in the revenues, expenses, and profits over which he or she has primary control.
- The income statement is naturally divided into three main parts:
  - Revenue
  - Expenses
  - Profit

Revenue Analysis

- In most cases, expenses for a business will go up continually. In the face of rising costs, you must increase your revenue levels to maintain or increase the amount of profit made by your business.
- When owners and managers of a restaurant seek to increase revenue, they must do so by:
  1. Increasing the number of guests served, and/or
  2. Increasing the average amount spent by each guest.
- In the hotel business, if management seeks to increase its rooms revenue, it must do so by:
  1. Increasing the number of rooms sold, and/or
  2. Increasing the average daily rate (ADR) for the rooms it sells.
- A careful analysis of revenue in a restaurant or hotel involves more than simply identifying the total amount of sales. A revenue performance of $50,000 could be viewed quite differently based upon a restaurant's prior performance (see Figure 3.10 for an example).
- In most cases, managers are concerned about the sources and total amount of sales or revenue their businesses have achieved, but they should also be interested in the changes (increases or decreases) to revenue experienced by their businesses.
- Increases in revenue, for example, may be tied to increased numbers of guests served, expansion of products and services sold, an increase in the number of hours the business is operated, or changes in the prices charged.
- Declines in revenue could indicate reduced numbers of guests served, reduced spending on the part of each guest served, or any number of a variety of other explanations that should be known by management.
- Managerial accountants must be very careful when analyzing changes in revenues. In some cases, revenues may have appeared to increase (or decrease) when in fact they have not. Additional factors that managerial accountants often consider when making a complete evaluation of revenue increases include:
  - The number of days included in the accounting period
  - Changes in the number of high or low volume days included in the accounting period

- o Differences in date placement of significant holidays (i.e. month or day of week)
- o Changes in selling prices
- o Variations in operational hours

Expense Analysis

- Guests cause businesses to incur costs. If there are fewer guests, there are likely to be fewer costs, but fewer profits as well!
- The real question to be considered is not whether costs are high or low. The question is whether costs are too high or too low, given management's view of the value it hopes to deliver to the guest and the goals of the operation's owners.

Profit (Loss) Analysis

- Many managers feel that it is the third (profit) section of the income statement that is the most important of all. Profit can be considered, to a large degree, the ultimate measure of the ability of hospitality professionals to plan and operate a successful business.
- The dollars which remain after all expenses have been paid represent the "profit" made by a business. Despite that fairly clear definition, profit is often viewed very differently by different readers of the income statement.
- To understand a significant reason why this is so, it is important for managers to understand the many factors that affect profit. For an example of this, see *Go Figure!* in your text.
- The profit section of an income statement most often will not be labeled as "profit" but rather will seek to tell the reader those factors that have, and have not, been included in the computation of net income.
- Net income, or profit, is sometimes known as the "**bottom line**", because it is often the "bottom-most" line on an income statement. In a properly prepared income statement, it represents the difference between all recorded revenue transactions and all recorded expense transactions.
- For most hospitality managers, the "bottom line" is *not* the most important number on the income statements they will generate. Hospitality managers (as opposed to the business's owners) may concern themselves most about income that remains after subtracting the expenses they can actually control.

Vertical Analysis

- To help managers carefully evaluate the revenues, expenses, and profit produced by their businesses, they can use a **vertical analysis** approach, which compares all items on the income statement to revenues using percentages. In this approach, an operation's total revenue figure takes a value of 100%.

- For example, when a hotel's accountant reports the costs of the hotel's food department, these are commonly expressed as a percentage of total hotel revenue (see Figure 3.11).

Go Figure!

When analyzing the Blue Lagoon Water Park Resort income statement as presented in Figure 3.11, managerial accountants would compute the food cost (food cost of sales) as a percentage of total revenue.

$$\frac{\text{Food Costs (Food Cost of Sales)}}{\text{Total Revenue}} = \text{Food Cost \%}$$

- When utilizing vertical analysis, individual sources of revenue and the operation's expenses are expressed as a fraction of total revenues. Each percentage computed is a percent of a "common" number. As a result, vertical analysis is also sometimes referred to as **common-size analysis**.
- All ratios are calculated as a percentage of total sales *except* the following:
  o Food costs are divided by food sales.
  o Beverage costs are divided by beverage sales.
  o Food gross profit is divided by food sales.
  o Beverage gross profit is divided by beverage sales.
- In restaurants, food and beverage items use their respective food and beverage sales as the denominator so that these items can be evaluated separately from total sales. Food and beverage costs are the most controllable items on the restaurant income statement.

Go Figure!

When analyzing Joshua's income statement as presented in Figure 3.12, managerial accountants would compute his food cost percentage as a percentage of "food" sales, rather than "total" sales.

$$\frac{\text{Food Costs}}{\text{Food Sales}} = \text{Food Cost \%}$$

- Food cost % for hotels is calculated by dividing food costs by *total* sales (revenue), whereas food cost % for restaurants is calculated by dividing food costs by *food* sales. Therefore, the authors of your text prefer the term "vertical analysis" to the term "common-size" analysis because the denominator is not always "common" among different industry segments.

- Vertical analysis may be used to compare a unit's percentages with industry averages, budgeted performance, other units in a corporation, or percentages from prior periods.
- **Profit margin** is the most telling indicator of a manager's overall effectiveness at generating revenues and controlling costs in line with forecasted results.

Go Figure!

As can be seen in Figure 3.12, profits for Joshua's refer to the net income figure at the bottom of his income statement. His profit percentage using the profit margin formula is as follows:

$$\frac{\text{Net Income}}{\text{Total Sales}} = \text{Profit margin}$$

- In both the USAR and USALI, an important objective is that of **responsibility accounting** for each separate department. It is important for upper management to know how each department is performing, so individual department managers can be held responsible for their own efforts and results.
- Expenses should increase when increased revenues require management to provide more products or labor to make the sale. If expenses are reduced too much (for example by choosing lower cost but inferior ingredients when making food for sale or providing too few front desk agents during a busy check-in period) the effect on revenue can be both negative and significant. Managers analyzing the expense portion of an income statement should be most concerned about the relationship between revenue and expense (vertical analysis) and less concerned about the total dollar amount of expense.

Go Figure!

The formula utilized to compute the specific expense percentage is:

$$\text{Specific Expense} / \text{Total Expenses} = \text{Specific Expense \%}$$

Horizontal Analysis

- Managers can utilize **horizontal analysis** (also called **comparative analysis**) to evaluate the dollars or percentage change in revenues, expenses, or profits experienced by a business.
- A horizontal analysis of income statements requires at least two different sets of data. Managers who use horizontal analysis to evaluate their income statements may be concerned with comparisons such as their:
  o Current period results vs. prior period results
  o Current period results vs. budgeted (planned) results

- o Current period results vs. the results of similar business units
- o Current period results vs. industry averages

Determining Variance

- The *dollar change* or **variance** shows changes from previously experienced levels, and will give you an indication of whether your numbers are improving, declining, or staying the same (see Figure 3.14).

Go Figure!

To calculate the variance, you would use the following formula:

> **Sales This Year – Sales Last Year = Variance**

Effective managers are also interested in computing the **percentage variance,** or percentage change, from one time period to the next. Thus, your sales percentage variance is determined as follows:

> **(Sales This Year – Sales Last Year)/Sales Last Year = Percentage Variance**

An alternative and shorter formula for computing the percentage variance is as follows:

> **Variance/ Sales Last Year = Percentage Variance**

Another way to compute the percentage variance is to use a math shortcut, as follows:

> **(Sales This Year/ Sales Last Year) –1 = Percentage Variance**

- All dollar variances and percentage variances of expenses on the income statement can be calculated in the same way. The dollar differences between identical categories listed on two different income statements is easy to compute and is always the numerator in any percentage change (variation) calculation. The denominator, however, varies.
- **When computing % change or variation in data from:**

When computing % change or variation in data from:	Use as the Denominator
Current Period to Prior Period	Prior Period's number
Current Period to Budget (Plan)	Budget (Plan) number
Current Period to Similar Business Unit	Similar Business Unit's number
Current Period to Industry Average	Industry Average number

- The comparative income statement helps managers analyze their expenses and take corrective action if it is needed (see Figure 3.16).

- There are two basic formulas used by managers to compare actual expenditures to budgeted expenditures. These are:
  - Variation (in dollars) from budget
  - Percent of variation from budget
- To compute the variation (in dollars) from budget, simply subtract the budgeted amount of expense from the amount actually spent.

Go Figure!

Thus, for example, (from Figure 3.16) in the area of *computer equipment*, the variation (in dollars) from budget for the Blue Lagoon's Property Operations and Maintenance department would be computed as:

$$\text{Actual Expense} - \text{Budgeted Expense} = \text{Variance}$$

Some managers prefer to express variations from budget in percentage terms. Thus, the percentage variance for computer equipment is determined as follows:

$$(\text{Actual Expense} - \text{Budgeted Expense}) / \text{Budgeted Expense} = \text{Percentage Variance}$$

An alternative and shorter formula for computing the percentage variance is as follows:

$$\text{Variance} / \text{Budgeted Expense} = \text{Percentage Variance}$$

Another way to compute the percentage variance is to use a math shortcut, as follows:

$$(\text{Actual Expense} / \text{Budgeted Expense}) - 1 = \text{Percentage Variance}$$

- In Figure 3.16, the computer equipment expense is "over" budget by $495, representing a relatively large percentage increase of 41.3%. Most managers would consider a 41.3% variation from budget to be very significant.
- On the other hand, as shown in Figure 3.16, the total property operations and maintenance expenses are $2,835 over budget, but the percentage variance is only 2.9%. In general, the larger the budgeted number, the smaller the percentage of variance from budget that will command significant management attention.
- If our budget was accurate, and we are within reasonable limits of our budget, we are said to be **in-line** or in compliance with our budget. If, as management, we decided that plus (more than) or minus (less than) a designated percentage (such as 10%) of budget in each category would be considered in-line or acceptable, we are in-line with regard to expenses.
- A significant variation is any variation in expected costs that management feels is a case for concern.
- Net income (profit) can be expressed on the income statement simply as a number or by using a form of either vertical analysis, horizontal analysis or both.

Managers seeking to make net income comparisons to budget or prior periods, either in terms of dollar differences or percentage differences will utilize the same mathematical formulas used to make revenue and expense comparisons.

Go Figure!

This is a skill you have already learned, as you can see from the following formulas based on the data in Figure 3.14!

> **Net Income This Year – Net Income Last Year = Variance**

Joshua's net income percentage variance is determined as follows:

> **(Net Income This Year – Net Income Last Year)/Net Income Last Year = Percentage Variance**

An alternative and shorter formula for computing the percentage variance is as follows:

> **Variance/Net Income Last Year = Percentage Variance**

Another way to compute the percentage variance is to use a math shortcut, as follows:

> **(Net Income This Year/Net Income Last Year) –1 = Percentage Variance**

Key Terms & Concepts Review

Match the key terms with their correct definitions.

1. Revenue _____ a. The last section of the USAR, which is least controllable by the foodservice manager and includes items such as interest and taxes.

2. Expenses _____ b. A company to whom a business owes money, such as a vendor.

3. Profit _____ c. Placement of expenses on the income statement.

4. Profit center _____ d. An analytical approach used to evaluate the dollars or percentage change in revenues, expenses, and profits experienced by a business requiring at least two different sets of data.

5. Business dining _____
6. Cost center _____
7. Stakeholders _____
8. Appreciation _____
9. Return on investment (ROI) _____
10. Debt _____
11. Equity _____
12. Terms _____
13. Interest _____
14. Default _____
15. Creditor _____
16. Invoice _____
17. Creditworthiness _____
18. Operate in the red _____
19. Operate in the black _____
20. Gross profit section (USAR) _____

e. Percentage change in revenues, expenses, and profits from one time period to the next.
f. Slang term used to describe a business that is not making a profit (losing money).
g. Refund on a bill.
h. An analytical approach which uses percentages to compare all items on the income statement to revenues.
i. The term used to indicate the dollars taken in by the business in a defined period of time. Often referred to as sales.
j. Funds supplied by investors or owners.
k. A phrase used to describe being within reasonable limits or in compliance with the budget.
l. Also referred to as Horizontal analysis.
m. An area in a business that generates revenues, expenses, and profits.
n. Individuals or companies directly affected by a business's profitability including owners, investors, lenders, creditors, and managers.
o. Also referred to as Departmental cost.
p. The first section of the USALI consisting of separate profit centers, which generate departmental income.
q. Decision made about how best to match revenues to expenses in an accounting period.
r. The difference between planned results and actual results.
s. The ability to pay bills promptly.
t. Food is provided as a service to the company's employees either as a no-cost (to the employee) benefit or at a greatly reduced price.

21. Operating expenses section (USAR) _____
22. Nonoperating expenses section (USAR) _____
23. Operated department income section (USALI) _____
24. Undistributed operating expenses section (USALI) _____
25. Nonoperating expenses section (USALI) _____
26. Timing _____
27. Classification _____
28. Invoice credit _____
29. Expense classification _____
30. Departmental cost _____
31. Direct operating expense _____
32. Undistributed operating expense _____

u. An approach to analyzing accounting information in which individual department managers are held responsible for their own efforts and results.

v. The second section of the USALI consisting of undistributed operating expenses, which are expenses that cannot truly be assigned to one specific department, and are thus, not distributed to the departments.

w. The first section of the USAR consisting of food and beverage sales, costs, and gross profits that can and should be controlled by the manager on a daily basis.

x. Slang used to describe net income or profit and refers to the bottom line of the income statement.

y. An accounting summary that closely examines the financial condition, or health, of a business by reporting the value of a company's total assets, liabilities, and owners' equity on a specified date.

z. The return on investment to a lender for funds lent (debt).

aa. The cost of the items required to operate the business.

bb. Funds lent to a business.

cc. Social, Military, Education, Religious, and Fraternal segments used to classify guests and the revenues they generate.

dd. A ratio of the money made compared to the money invested.

ee. Also referred to as Vertical analysis.

ff. The percentage of net income to revenues.

33. Schedules _____ gg. A bill from a vendor detailing the purchases made by a business.

34. Package _____ hh. The last section of the USALI, which is least controllable by the manager and includes items such as interest and taxes.

35. SMERF _____ ii. Increase in property value.

36. Bottom line _____ jj. The second section of the USAR, which contains operating expenses controllable by the manager on a weekly or monthly basis (with the exception of wages, which can be controlled daily).

37. Vertical analysis _____ kk. An expense that is attributable to one department.

38. Common-size analysis _____ ll. The dollars that remain after all expenses have been paid.

39. Profit margin _____ mm. Tools used by managerial accountants to provide statement readers with more in-depth information about important areas of revenues and expenses.

40. Responsibility accounting _____ nn. The process of carefully considering how a business's expenses will be detailed for reporting purposes.

41. Horizontal analysis _____ oo. A specially packaged collection of goods and services.

42. Comparative analysis _____ pp. An expense that cannot truly be assigned to one specific area within an operation.

43. Variance _____ qq. Slang term used to describe a business that is profitable.

44. Percentage variance _____ rr. Conditions of a loan.

45. In-line _____ ss. A unit that generates costs but no revenue or profits.

46. Balance sheet _____ tt. Non-payment of a loan.

Discussion Questions

1. List the five stakeholders affected by a business's profitability.
_____
_____
_____

2. List the three sections of a USAR income statement.

3. List the three sections of a USALI income statement.

4. Describe the two ways an income statement (as a whole) is normally analyzed.

Quiz Yourself
Choose the letter of the best answer to the questions listed below.

1. John's brother has invested $50,000 in his restaurant. In its first year, the restaurant had revenues of $250,000, and John's brother received $5,000 on his investment. What is his ROI?
   a. 10%
   b. 20%
   c. 2%
   d. 50%

2. A summary of financial information for a defined accounting period is called a(n)
   a. Balance sheet
   b. Funding statement
   c. Income statement
   d. Statement of revenues and expenses

3. In evaluating the revenues, expenses and profit produced by their business, managers can compare all items to revenues using percentages. This method is called
   a. Profit/loss analysis
   b. Vertical analysis
   c. Percentage analysis
   d. Revenue analysis

4. Last year, the Bluebonnet Inn, a 6 bedroom Bed & Breakfast, had total revenues of $207,700. The Inn had total food costs (food cost of sales) of $18,700 and total food sales of $62,500. What is the Inn's food cost percentage?
   a. 12.6%
   b. 30.1%
   c. 29.9%
   d. 9.0%

5. Ellington's Bar & Grill had total sales of $346,700 this year, including $249,600 in total food sales. Total food costs for the year are $79,900. Calculate the food cost percentage for the year.
   a. 32.0%
   b. 23.0%
   c. 60.5%
   d. 43.6%

6. Ellington's Bar & Grill had a net income of $32,580 this year, with total sales of $346,700 and total gross profit of $222,200. Calculate the profit margin for the year.
   a. 14.7%
   b. 10.6%
   c. 9.4%
   d. 64.1%

7. If Ellington's Bar & Grill spent $2,249 on newspaper advertising, with a total marketing expense of $8,650, and total operating expenses of $161,700, what is the specific expense percentage for advertising?
   a. 1.4%
   b. 26.0%
   c. 67.3%
   d. 7.0%

8. If sales last year were $336,550 and sales this year are $346,700, what is the sales variance?
   a. $ 10,150
   b. ($10,150)
   c. $ 103
   d. $ 97

9. If the budgeted amount for equipment repairs was $10,300, and the actual amount spent on this category was $9,850, what is the percentage variance?
   a. 450.0%
   b. 45.0%
   c. (4.4%)
   d. 4.6%

10. Ellington's Bar & Grill had budgeted a net income for this year of $30,960, but the actual net income for the year is $32,580. Last year, actual net income was $30,150. What is the percentage variance in net income?
    a. 92.5%
    b. 8.1%
    c. 10.8%
    d. 7.5%

Chapter 4

The Balance Sheet

Highlights

* The Purpose of the Balance Sheet
* Balance Sheet Formats
* Balance Sheet Content
* Balance Sheet Analysis
* Key Terms and Concepts Review
* Discussion Questions
* Quiz Yourself

Study Notes

The Purpose of the Balance Sheet

- Business owners prepare balance sheets to better understand the value of a business and how well its assets have been utilized to produce wealth for the business's owners. For an example of why this is important, see *Go Figure!* in the text.
- The type of information contained on a business's balance sheet is of critical importance to several different groups including:
  - Owners
  - Investors
  - Lenders
  - Creditors
  - Managers

Owners

- The balance sheet, prepared at the end of each defined accounting period, lets the owners of the business know about the amount of that business which they actually "own".
- A **lien** is the legal right to hold another's property to satisfy a debt. A bank's lien is similar to a business's liabilities. These liabilities must be subtracted from the value of the business before its owners can determine the amount of their own equity (free and clear ownership). The balance sheet is designed to show the amount of a business owner's free and clear ownership.

Investors

- Investors seek to maximize the return on investment (ROI) (see Chapter 1) they receive. When a business's balance sheet from one accounting period is compared to its balance sheet covering another time period, investors can measure their return on investment. For an example of this, see *Go Figure!* in the text.
- Investors must have the information contained in a balance sheet if they are to accurately compute their annual returns on investment (ROI).

Lenders

- Lenders are most concerned about a business's ability to repay its debts. For a comparison of two hotels' abilities to repay their debts, see *Go Figure!* in the text.
- Lenders read the balance sheet of a business in an effort to better understand the financial strength (and thus the repayment ability) of that business.

Creditors

- Business creditors, much like lenders, are concerned about repayment. It would not be unreasonable for vendors to ask to see their customer's respective balance sheets before a decision was made regarding the wisdom of extending credit to them. The balance sheet is the financial document that most accurately indicates the long-term ability of a business to repay a vendor who has extended credit to that business.

Managers

- Managers most often are more interested in the information found on the income statement than that found on the balance sheet. However, they too must be able to read and analyze their own balance sheets to determine items such as the current financial balances of cash, accounts receivable, inventories, and accounts payable, and other accounts that have a direct impact on operations.
- Those who operate a business must consistently have sufficient cash on hand to pay their employees, their vendors, and the taxes owed by the business. The balance sheet indicates to management the amount of cash available to them on the day the balance sheet is prepared.
- Managers must be able to read a balance sheet to determine the total amount of accounts receivable owed to the business on the date it is prepared. Excessively large amounts of accounts receivable (which are not identified on the income statement) could be telltale signs that:
    - Too much credit has been extended
    - Credit collection efforts may need to be reviewed and improved if necessary
    - Cash reserves could become insufficient to meet the short term needs of your business

Limitations of the Balance Sheet

- The balance sheet is important because it reveals, at a fixed point in time, the amount of **wealth** that a company possesses. In this case, wealth is defined as the current value of all a company's assets minus all of the company's obligations.
- No single approach to valuing assets is used by accountants in the preparation of the balance sheet. Knowledgeable readers of a balance sheet recognize that accountants utilize a variety of evaluation approaches, each of which may make the most sense for specific asset types based upon circumstances and available information.
- It is also important to note that balance sheets have been criticized because of the company assets they do *not* value. Consider the fact that, of all the assets listed on the balance sheet, none take into account the relative value, or worth, of a restaurant or hotel's staff, including its managers.

Balance Sheet Formats

- A balance sheet represents an accountant's systematic method of documenting the value of a business's assets, liabilities and owner's equity on a specific date.
- There are two basic methods accountants use to display the information on a balance sheet.
- When using the **account format** those preparing the balance sheet list the assets of a company on the left side of the report and the liabilities and owner's equity accounts on the right side (see Figure 4.1).
- When using the **report format** those preparing the balance sheet list the assets of a company first and then the liabilities and owner's equity accounts (vertically), and present the totals in such a manner as to prove to the reader that assets equals liabilities plus owners equity (see Figure 4.2).
- In both type formats, the date on which the balance sheet was prepared is clearly identified.

Balance Sheet Content

- The Accounting formula is stated as *Assets = Liabilities + Owner's Equity*, and the purpose of a balance sheet is to tell its readers as much as possible about each of these three accounting formula components.

Why a Balance Sheet Balances - Dr. Dopson's Stuff Theory

- Dr. Lea Dopson, a hospitality managerial accounting professor and one of the authors of your textbook, has developed "Dr. Dopson's Stuff Theory" to explain why a balance sheet balances (see Figure 4.3).
- Dr. Dopson's Stuff Theory is simple. For all the stuff (assets) you have in your life, you got it from either:

- o Borrowing money that you have to repay such as through credit cards or loans (liabilities) or
- o Acquiring the stuff through others such as your parents, siblings, or friends (investors' equity) or paying for the stuff yourself through money you earned (retained earnings equity)
- You are reporting your stuff and how you GOT (past tense) your stuff. In this sense, the list of your stuff *balances* with how you got it. This is why a balance sheet balances!
- This same concept applies to businesses when preparing their balance sheets. All assets (stuff) must equal liabilities plus owners' equity (how they got their stuff).

Components of the Balance Sheet

- The balance sheet is often subdivided into components under the broad headings of Assets, Liabilities, and Owners' Equity.
- These subclassifications have been created by accountants to make information more easily accessible to readers of the balance sheet and to allow for more rapid identification of specific types of information for decision making.

Assets

Current Assets

- **Current Assets** are those which may reasonably be expected to be sold or turned into cash within one year (or one operating season).
- **Liquidity** is defined as the ease in which current assets can be converted to cash in a short period of time (less than 12 months).
- Current assets, typically listed on the balance sheet in order of their liquidity, include:
  - o Cash
  - o Marketable securities
  - o Accounts receivable (net receivables)
  - o Inventories
  - o Prepaid expenses
- For purposes of preparing a balance sheet, the term **cash** refers to the cash held in cash banks, money held in checking or savings accounts, electronic fund transfers from payment card companies, and **certificates of deposit (CDs).**
- **Marketable securities** include those investments such as stocks and bonds that can readily be bought sold and thus are easily converted to cash. These are stocks and bonds the business purchases from *other* companies. These are not to be confused with a company's stocks that are listed on its balance sheet as owners' equity.
- Accounts receivable represent the amount of money owed to a business *by others* (such as customers). *Net* **receivables** (the term *net* means that something has been

subtracted out) are those monies owed to the business after subtracting any amounts that may not be collectable (**doubtful accounts**).
- In the hospitality industry, **inventories** will include the value of food, beverages and supplies used by a restaurant, as well as sheets, towels and the in-room replacement items (hangers, blow dryers, coffee makers and the like) used by a hotel.
- **Prepaid expenses** are best understood as items that will be used within a year's time, but which must be completely paid for at the time of purchase.
- The order of liquidity for current assets is easily explained:
    - Cash is listed first because it is already cash.
    - Marketable securities are less liquid than cash, but can be readily sold for cash.
    - Net receivables can be collected from others (customers), but not as easily as converting marketable securities to cash.
    - Inventories must be made ready for sale to customers and the money must be collected. There is no guarantee that payment for all inventories will be collected in full, and thus some may end up being reported as receivables.
    - Prepaid expenses are the least liquid current asset because once paid, refunds for this money are very difficult (if not impossible) to receive. For an illustration of prepaid expenses, see *Go Figure!* in the text.
- Those prepaid expenses that will be of value or benefit to the business for more than one year (for example a three year pre-paid insurance policy) should be listed on the balance sheet as "Other Assets."

Non Current (Fixed) Assets

- **Non Current (Fixed) Assets** consist of those assets which management intends to keep for a period longer than one year, and typically include investments, property and equipment (land, building, furnishings and equipment, less accumulated depreciation), and other assets (see Figures 4.1 and 4.2).
- Included in this group are **investments** made by the business which management intends to retain for a period of time longer than one year, unlike marketable securities, which can be readily sold and converted to cash within one year.
- Investments are typically one of three types:
    - **Securities** (stocks and bonds) acquired for a specific purpose, such as a restaurant company that purchases a significant amount of stock in a smaller company with the intent of influencing the operations of the smaller company.
    - Assets owned by a business but not currently used by it. An example would be vacant land owned by a restaurant company that is going to build (but has not yet built) a restaurant on the site.
    - Special funds that have a specific purpose. The most common of these is a **sinking fund**, in which monies are reserved and invested for use in the future.
- Investments are valued on the balance sheet at their fair market value. **Fair market value** is most often defined as the price at which an item would change hands between a buyer and a seller without any compulsion to buy or sell.

- **Property and Equipment**, which includes land, building, furnishings and equipment, usually make up a significant portion of the total value of a hospitality business and are another form of non-current asset. These are listed on the balance sheet at their original cost less their accumulated depreciation (see Chapter 2).
- **Other assets** are a non-current asset group that includes items that are mostly intangible. This includes the value of **goodwill** (the difference between the purchase price of an item and its fair market value). Goodwill is an asset that is **amortized** (systematically reduced in value) over the period of time in which it will be of benefit to the business.
- Other items in this category may include the cash surrender value of life insurance policies taken out on certain company executives, deferred charges such as loan fees that are amortized over the life of the loan, and restricted cash that is set aside for a specific (restricted) purpose.
- The techniques of depreciation and amortization used by a business should be prepared and shown in the Notes to Financial Statements that should be attached to the balance sheet.
- As previously noted it can be difficult to accurately assess the value of a business's assets. The following approaches to evaluation are typically used by hospitality accountants when they seek to establish the value of a company's assets:

Asset Type	Worth Established By
Cash	Current value
Marketable securities	Fair market value or amortized cost
Accounts receivable	Estimated future value
Inventories	The lesser of current value or price paid
Investments	Fair market value or amortized cost
Property and equipment	Price paid adjusted for allowed depreciation.

Liabilities

- **Current liabilities** are defined as those obligations of the business that will be repaid within a year. The most important sub-classifications of current liabilities include notes payable, income taxes payable, and accounts payable.
- In the hospitality industry, current liabilities typically consist of payables resulting from the purchase of food, beverages, products, services and labor.
- Current period payments utilized for the reduction of long-term debt are also considered a current liability.
- Guest's pre-paid deposits (such as those required by a hotel that reserves a ballroom for a wedding to be held several months in the future) are also listed as a current liability (because these monies are held by the business but have not been earned at the time the hotel accepts the deposit).
- Dividends that have been declared but not yet paid and income taxes that are due but not yet paid will also be included in this liability classification.
- **Long-term liabilities** are those obligations of the business that will not be completely paid within the current year. Typical examples include long-term debt,

mortgages, lease obligations, and deferred incomes taxes resulting from the depreciation methodology used by the business.

Owner's Equity

- While the liabilities section of the balance sheet identifies non-owner (external) claims against the business's assets, the owners' equity portion identifies the asset claims of the business's owners.
- For corporations, **common stock** is the balance sheet entry that represents the number of shares of stock issued (owned) multiplied by the **par value** (the value of the stock recorded in the company's books).
- Common stock is valued at its historical cost regardless of its current selling price. Initially, most companies designate a stated or par value for the stock they issue and as each share is sold, an amount equal to the par value is reported in the common stock section of the balance sheet.
- Differences between the selling price and par value are reported in the **paid in capital** portion of the balance sheet.
- Some companies also issue **preferred stock** that pays its stockholders (owners) a fixed dividend. When more than one type of stock is issued by a company, the value of each type should be listed separately on the balance sheet.
- Retained earnings are the final entry on the owners' equity portion of the balance sheet. **Retained earnings** refer simply to the accumulated account of profits over the life of the business that have not been distributed as dividends. If net losses have occurred, the entry amount in this section may be a negative number.
- When a company is organized as a sole proprietorship, the balance sheet reflects that single ownership (see Figure 4.6). When a partnership operates the business, each partner's share is listed on the balance sheet (see Figure 4.7).

Balance Sheet Analysis

- After the balance sheet has been prepared, managerial accountants use a variety of methods to analyze the information it contains. Three of the most common types of analysis are:
  o Vertical (common-size) analysis
  o Horizontal (comparative) analysis
  o Ratio analysis

Vertical Analysis of Balance Sheets

- Income statements can be analyzed using vertical and horizontal techniques. In very similar ways, the balance sheet can be reviewed using these same techniques.
- When using vertical analysis on a balance sheet, the business's Total Assets take on a value of 100%. Total Liabilities and Owners' Equity also take a value 100% (see Figure 4.8).

- When utilizing vertical analysis, individual asset categories are expressed as a percentage (fraction) of Total Assets. Individual liability and owner's equity classifications are expressed as a percentage of Total Liabilities and Owner's Equity.
- Each percentage computed is a percent of a "common" number, which is why this type analysis is sometimes referred to as common-size analysis.
- Vertical analysis of the balance sheet may be used to compare a unit's percentages with industry averages, other units in a corporation, or percentages from prior periods.

Horizontal Analysis of Balance Sheets

- A second popular method of evaluating balance sheet information is the horizontal analysis method.
- As was true with the income statement, a horizontal analysis of a balance sheet requires at least two different sets of data. Because of this "comparison" approach the horizontal analysis technique is also called a comparative analysis.
- Managers who use horizontal analysis to evaluate the balance sheet may be concerned with comparisons such as their:
  - Current period results vs. prior period results (see Figure 4.9)
  - Current period results vs. budgeted (planned) results
  - Current period results vs. the results of similar business units
  - Current period results vs. industry averages

Determining Variance

- The *dollar change* or variance shows changes from previously experienced levels, and will give you an indication of whether your numbers are improving, declining, or staying the same (see Figure 4.9).

Go Figure!

To calculate the variance in cash, you would use the following formula:

Cash This Year – Cash Last Year = Variance

The cash percentage variance is determined as follows:

(Cash This Year – Cash Last Year)/Cash Last Year = Percentage Variance

An alternative and shorter formula for computing the percentage variance is as follows:

Variance/Cash Last Year = Percentage Variance

Another way to compute the percentage variance is to use a math shortcut, as follows:

> **(Cash This Year/Cash Last Year) −1 = Percentage Variance**

- All dollar variances and percentage variances on the balance sheet can be calculated in the same way. The dollar differences between identical categories listed on two different balance sheets is easy to compute and is always the numerator in any percentage change (variation) calculation. The denominator, however, varies.
- **When computing % change or variation in data from:**

When computing % change or variation in data from:	Use as the Denominator
Current Period to Prior Period	Prior Period's number
Current Period to Budget (Plan)	Budget (Plan) number
Current Period to Similar Business Unit	Similar Business Unit's number
Current Period to Industry Average	Industry Average number

- Managerial accountants reviewing financial data are often concerned about *both* dollar change and percentage change because a dollar change may at first appear large, but when compared to its base figure, represent a very small percentage change.
- To illustrate, consider the "Other Assets" and "Total Assets" sections of Figure 4.9. The percentage change in Other Assets is 13.8%, which reflects a dollar change of $81,000. However, the percentage change in "Total Assets" is actually smaller at 2.2%, but represents a much larger dollar change of $988,945.

Inflation Accounting

- The second reason for computing percentage change in the preparation of balance sheets (as well as other financial documents) is because the value of money changes over time.
- As economists (as well as managerial accountants) know well, if you don't adjust for **inflation** (the tendency for prices and costs to increase) just about *everything* is more expensive today than it was 30 years ago.
- Because of the potential effects of inflation and **deflation** (the tendency for prices and costs to decrease), managerial accountants must know how to properly consider them when comparing financial data from two different time periods.
- The **purchasing power** (amount of goods and service that can be bought) of money is likely diminished over time due to the effects of inflation.
- Accounting for inflation (sometimes called **current dollar accounting**) can be very complex. A managerial accountant can easily understand the major issues addressed by inflation accounting and use basic techniques that will help address these issues.

Key Terms & Concepts Review

Match the key terms with their correct definitions.

1. Lien _____
2. Wealth _____
3. Account format (balance sheet) _____
4. Report format (balance sheet) _____
5. Current assets _____
6. Liquidity _____
7. Cash _____
8. Certificates of deposit (CDs) _____
9. Marketable securities _____
10. Net receivables _____
11. Doubtful accounts _____
12. Inventories _____
13. Prepaid expenses _____
14. Non current (fixed) assets _____
15. Investments _____

a. Assets that are intended to be retained for a period of time longer than one year such as land, buildings, furnishings, and equipment.

b. The amount of money that may not be collectable from receivables.

c. The value of the food, beverages, and supplies used by a restaurant, as well as sheets, towels, and in-room replacement items used by a hotel.

d. The portion of the balance sheet that reports any differences between the selling price and par value of stock.

e. Accounting for inflation.

f. Stocks or bonds.

g. Assets which management intends to keep for a period longer than one year including the property building(s), and equipment owned by a business.

h. Stock that provides a fixed dividend to stockholders.

i. The legal right to hold another's property to satisfy a debt.

j. The value of stock recorded in the company's books.

k. A balance sheet format that lists the assets of a company and the liabilities and owners' equity accounts (vertically), and presents the totals in such a manner as to prove to the reader that assets equals liabilities plus owners equity.

l. The money held in cash banks, checking or savings accounts, electronic fund transfers from payment card companies, and certificates of deposit.

m. The difference between the purchase price of an item and its fair market value.

n. Assets that are intended to be retained for a period of time longer than one year including items that are mostly intangible.

o. Those obligations of a business that will not be completely paid within the current year.

16. Securities   _____   p. Systematically reduced in value.

17. Sinking fund   _____   q. A balance sheet format that lists the assets of a company on the left side of the report and the liabilities and owners' equity accounts on the right side of the report.

18. Fair market value   _____   r. Assets which may reasonably be expected to be sold or turned into cash within one year (or one operating season).

19. Property and equipment   _____   s. Accumulated account of profits over the life of the business that have not been distributed as dividends.

20. Other assets   _____   t. Amount of goods and service that can be bought.

21. Goodwill   _____   u. The current value of all a company's assets minus all of a company's obligations.

22. Amortized   _____   v. Money owed by customers to a business calculated after subtracting any amounts that may not be collectable.

23. Current liabilities _____   w. Financial instruments with a fixed term and interest rate and are considered cash because of their liquidity.

24. Long-term liabilities _____ x. The balance sheet entry that represents the number of shares of stock issued (owned) multiplied by the value of each share.

25. Common stock   _____   y. The price at which an item would change hands between a buyer and a seller without any compulsion to buy or sell, and with both having reasonable knowledge of the relevant facts.

26. Par value   _____   z. Assets that are intended to be retained for a period of time longer than one year such as a security (stock or bond), asset owned by a business but not currently used by it, and a special fund such as a sinking fund.

27. Paid in capital   _____   aa. Money that is reserved and invested for use in the future.

28. Preferred stock   _____   bb. Those obligations of a business that will be repaid within a year.

29. Retained earnings   _____   cc. Investments such as stocks or bonds that can readily be bought sold and thus are easily converted to cash.

30. Inflation   _____   dd. The ease in which current assets can be converted to cash in a short period of time (less than 12 months).

31. Deflation   _____   ee. The tendency for prices and costs to increase.

32. Purchasing power _____ ff. Item that will be used within a year's time, but which must be completely paid for at the time of purchase.

33. Current dollar accounting _____ gg. The tendency for prices and costs to decrease.

Discussion Questions

1. List the five groups affected by the information contained in the balance sheet and briefly discuss what they are looking for on the balance sheet.
   _____
   _____
   _____

2. Explain why the balance sheet balances.
   _____
   _____
   _____

3. List and briefly explain the three major components of the balance sheet.
   _____
   _____
   _____

4. Describe the two ways a balance sheet (as a whole) is normally analyzed.
   _____
   _____
   _____

Quiz Yourself

Choose the letter of the best answer to the questions listed below.

1. An accountant's systematic method of documenting the value of a business's assets, liabilities and owner's equity on a specific date is called a(n)
   a. Owners' report
   b. Balance sheet
   c. Income statement
   d. Retained earnings statement

2. The Accounting Formula is stated as
   a. Assets + Owners' Equity = Liabilities
   b. Assets = Owners' Equity - Liabilities
   c. Assets + Liabilities = Owners' Equity
   d. Assets = Liabilities + Owners' Equity

3. Assets which may be expected to be sold or converted to cash within one year are called
   a. Current assets
   b. Current liabilities
   c. Owners' equity
   d. Fixed assets

4. The value of items such as sheets, towels, and blow dryers is reported as
   a. Cash
   b. Marketable securities
   c. Inventories
   d. Fixed assets

5. Assets which management intends to retain for more than one year are called
   a. Current assets
   b. Fixed assets
   c. Inventories
   d. Owners' equity

6. Obligations of a business that will be repaid within a year are known as
   a. Investments
   b. Marketable securities
   c. Current liabilities
   d. Current assets

7. Retained earnings are described as
   a. Dividends on preferred stock
   b. Accumulated dividends that have been paid to owners
   c. Accumulated dividends that owners have reinvested into the business
   d. Accumulated profits that have not been distributed as dividends

8. The balance sheet analysis method that compares a business's percentages with industry averages or percentages from other units in a corporation is called
   a. Vertical analysis
   b. Revenue analysis
   c. Expense analysis
   d. Profit/loss analysis

9. Dollar variance in an asset account from a prior period can be computed using the following formula:
   a. Asset value this year – asset value last year
   b. Asset value this year / asset value last year
   c. Asset value this year / budgeted asset value
   d. Variance / asset value this year

10. The tendency for prices and costs to increase over time is called
    a. Purchasing power
    b. Inflation
    c. Deflation
    d. Step costs

Chapter 5

The Statement of Cash Flows

Highlights

* Understanding Cash Flows
* The Purpose of the Statement of Cash Flows (SCF)
* Sources and Uses of Funds
* Creating the Statement of Cash Flows
* Statement of Cash Flows Analysis
* Key Terms and Concepts Review
* Discussion Questions
* Quiz Yourself

Study Notes

Understanding Cash Flows

- Assume, for example, that you have income from a variety of sources. These sources may include money paid to you from a job, income from parents or other family members, and interest you may earn on savings accounts.
- In addition, you have living and school expenses that must be paid. You know that you must have enough money on hand to pay your bills as they become due. You will also find that if you do not have sufficient cash to pay your bills, it will cost you *more* to pay those bills than it otherwise would have. For an illustration of this concept, see *Go Figure!* in the text.
- Having access to cash at the right time (versus having the right amount of cash) is important to individuals. It is also critically important to businesses.

The Purpose of the Statement of Cash Flows (SCF)

- For many years, businesses that issued income statements (see Chapter 3) and balance sheets (see Chapter 4) to those outside the company were encouraged to also supply a document called the **statement of changes in financial position**, or funds statement. The intent of this statement was to indicate how cash inflows and outflows affected the business during a specific accounting period.
- In 1971 the Accounting Principles Board (APB) officially made the funds statement one of the three primary financial documents required in annual reports to shareholders. In late 1987, the Financial Accounting Standards Board (FASB) called for a statement of cash flows to replace the more general funds statement.

- The **statement of cash flow shows (SCF)** shows all sources and uses of funds from operating, investing, and financing activities of a business.
- A **consolidated income statement** combines the revenue, expense, and profit information from each individual sub-statement of a company. This can be used as a resource for developing a statement of cash flows.
- The cash inflows and outflows of a business are of significant importance to a business's owners, investors, lenders, creditors, and managers. The presentation of accurate information about cash flows should enable investors in a business to:
    o Predict the amount of cash that can be distributed via profit payouts or dividend distributions
    o Evaluate the possible risk associated with a business
- **Solvency** refers to the ability of a business to pay its debts as they become due. Solvency is an important measure of a firm's likelihood to remain a going concern (a business that generates enough cash to stay in business).
- If a business is not considered by lenders and investors to be a going concern, that business will likely find that its ability to borrow money is severely diminished. If it can borrow money, it will find that the money it does borrow will come with the increased costs that reflect the higher risk associated with lending money to a business that does not consistently demonstrate a strong positive cash flow.
- When cash flows are not positive, investors will demand higher return on investment (ROI) levels to compensate for the greater risk they are taking, and lenders will seek higher interest rates to compensate themselves for that same higher risk level.
- Cash on a balance sheet is considered to be a current asset. **Cash,** in this case, refers to currency, checks on hand, and deposits in banks. Cash is not synonymous with the term "revenue" or "sales."
- **Cash equivalents** are short-term, temporary investments such as treasury bills, certificates of deposit, or commercial paper that can be quickly and easily converted to cash.
- The SCF is designed to report a business's **sources and uses of funds** (inflows and outflows of money affecting the cash position) as well as its beginning and ending cash and cash equivalents balances for each accounting period.

Sources and Uses of Funds

- Sources represent inflows and uses represent outflows of funds for the hospitality business.
- When comparing assets from last period's balance sheet to this period's balance sheet, decreases in assets represent sources of funds and increases in assets represent uses of funds (see Figure 5.1).
- Accumulated depreciation behaves in an opposite manner to the other assets. This is because depreciation is a contra asset account (see Chapter 2).
- The direct effect of increasing or decreasing each asset account, with examples, is shown on pages 148-150 in the textbook.

- Alternatively, increases in liabilities and owners' equity represent sources of funds and decreases in liabilities and owners' equity represent uses of funds (see Figure 5.2).
- The direct effect of increasing or decreasing each asset account, with examples, is shown on pages 151-152 in the textbook.
- There is a trick to help you remember all of this as shown in Figure 5.3.

Sources	Uses
↓ Assets*	↑ Assets*
↑ Liabilities	↓ Liabilities
↑ Owners' Equity	↓ Owners' Equity

   * Remember that depreciation is a contra asset account and behaves oppositely of all other assets, so ↑ in depreciation is a source and ↓ of depreciation is a use.

- Up arrows represent "increases" and down arrows represent "decreases." Assets in each column have opposite arrows from liabilities and owners' equity. Also, arrows in the left column are opposite of those in the right column. If you can remember only one, you can remember all the rest. Once you have that arrow correct, then you can remember the directions of the other arrows. Be careful, though! If you get your one example backwards, then ALL of the others are wrong! Memorize this "trick"; it will help you immensely!
- An example of sources and uses of funds with numbers is found in Figure 5.4.

Creating the Statement of Cash Flows

- The statement of cash flows should be prepared just as often as you prepare your income statement and balance sheet. In order to build a statement of cash flows, you will need the following:
    - Income statement for this year, including a statement of retained earnings
    - Balance sheet from last year
    - Balance sheet from this year
- A **condensed income statement** (see Figure 5.10) for a hotel reports the revenues, expenses, and profits in a summary format, absent of specific departmental and undistributed expense details.
- A **statement of retained earnings** reports the changes in retained earnings (accumulated account of profits over the life of the business that have not been distributed as dividends) from last year to this year.
- Last year's balance sheet and this year's balance sheet are needed to report the changes in balance sheet accounts from one year to the next. This will show the sources and uses of funds that will affect the cash changes reported on the statement of cash flows (see Figure 5.11).
- The format of a SCF consists of the following:
    - Cash flow from operating activities

- o Cash flow from investing activities
- o Cash flow from financing activities
- o Net changes in cash
- o Supplementary schedules
- Figure 5.5 is an example of the standard format used to prepare a statement of cash flows.

Cash Flow from Operating Activities

- **Cash flow from operating activities** is the result of all of the transactions and events that normally make up a business's day to day activities. These include cash generated from selling goods or providing services, as well as income from items such as interest and dividends. Operating activities will also include cash payments for items such as inventory, payroll, taxes, interest, utilities, and rent.
- The net amount of cash provided (or used) by operating activities is a key figure on a statement of cash flows because it shows cash flows that managers can control the most.
- The first step in creating a statement of cash flows is to develop a summary of cash inflows and outflows resulting from operating activities (see Figure 5.6), using information provided on the income statement including sales, expenses, and thus, net income.
- There are two methods that are used in calculating and reporting the amount of cash flow from operating activities on the statement of cash flows: the indirect method and the direct method. Although both produce identical results, the indirect method is more popular because it more *easily* reconciles the difference between net income and the net cash flow provided by operations.
- When using the indirect method, you start with the figure for net income (taken from your income statement) and then adjust this amount up or down to account for any income statement entries that do not actually provide or use cash.
- The accrual income statement must be converted to a cash basis in order to report cash flow from operating activities.
- The two most common items on the income statement that may need to be adjusted from an accrual basis to a cash basis are
  - o Depreciation, and
  - o Gains/losses from a sale of investments/equipment.
- Depreciation is a method of allocating the cost of a fixed asset over the useful life of the asset (see Chapter 2). More important, however, depreciation is subtracted from the income statement primarily to lower income, and thus lower taxes. In order to adjust net income to reflect actual cash, then, depreciation must be added back as in Figure 5.12.
- A gain on a sale of an investment/equipment occurs when the original cost of the investment/equipment is lower than the price at which it is sold at a later date. Conversely, a loss on a sale of an investment/equipment occurs when the original cost of the investment/equipment is higher than the price at which it is sold at a later date. For an illustration of the effects of a sale of a fixed asset, see *Go Figure!* in the text.

- The remaining adjustments to net income when calculating cash flow from operating activities come from the sources and uses of funds calculated from the balance sheets. *Sources of funds are shown as a positive number on the statement of cash flows and uses of funds are shown as a negative number of the statement of cash flows.*
- In general, the sources and uses of funds used for cash flow from operating activities will come from current assets and current liabilities. The exceptions to this are marketable securities, which belongs in investing activities, and notes payable (short-term debt), which belongs in financing activities.

Cash Flow from Investing Activities

- An investment (see Chapter 4) can be understood simply as the acquisition of an asset for the purpose of increasing future financial return or benefits. **Cash flow from investing activities** summarizes this part of a business's action.
- A business's investing activities include those transactions and events involving the purchase and sale of marketable securities, investments, land, buildings, equipment, and other assets not generally purchased for resale (see Figure 5.7).
- The cash flow from investing activities comes from the sources and uses of funds that you calculated from your balance sheets. *Sources of funds are shown as a positive number on the statement of cash flows and uses of funds are shown as a negative number of the statement of cash flows.*
- In general, the sources and uses of funds used for cash flow from investing activities will come from long-term assets (investments, property and equipment, and other assets). The exception to this is marketable securities, which is a current asset that belongs in investing activities.

Cash Flow from Financing Activities

- The third and final of the three cash inflow and outflow activity summaries that make up a complete SCF relates to the financing activities of a business. **Cash flow from financing activities** refers to a variety actions including:
  o Obtaining resources (funds) from the owners of a business (e.g. by selling company stocks)
  o Providing owners with a return of their original investment amount (e.g. payment of dividends)
  o Borrowing money
  o Repaying borrowed money
- Although repayments of loans are considered a financing activity, interest paid and interest received are classified as operating activities (as part of the income statement).
- Cash payments made to reduce the **principal** (the amount borrowed) of a loan would be considered cash flow related to a financing activity, while any interest paid to secure the loan would be considered an operating expense.
- Loans, notes, and mortgages are all examples of financing activities that affect cash flows (see Figure 5.8).

- The cash flow from financing activities comes from the sources and uses of funds calculated from the balance sheets. *Sources of funds are shown as a positive number on the statement of cash flows and uses of funds are shown as a negative number of the statement of cash flows.*
- In general, the sources and uses of funds needed for cash flow from financing activities come from long-term debt and equity. The exception to this is notes payable (short-term debt), which is a current liability that belongs in financing activities. Also, dividends paid must be recorded here because that is a cash outflow from net income.
- Additions and subtractions to the statement of cash flows are shown in Figure 5.9. With the exceptions noted, operating activities are developed using current assets and current liabilities, investing activities are developed using long-term assets, and financing activities are developed using long-term debt and owners' equity.

Operating activities
Net income
+/- Depreciation
+/- Losses/gains from the sale of investments/equipment
+/- Current assets (except marketable securities)
+/- Current liabilities (except notes payable)

Investing activities
+/- Marketable securities
+/- Investments
+/- Property and equipment
+/- Other assets

Financing activities
+/- Notes payable
+/- Long-term debt
+/- Common stocks and paid in capital
+/- Dividends paid

Net Changes in Cash

- **Net changes in cash** represent all cash inflows minus cash outflows from operating, investing, and financing activities. This net change in cash must equal the difference between the cash account at the beginning of the accounting period and the cash account at the end of the accounting period. This is illustrated in *Go Figure!* in the text.

Supplementary Schedules

- **Supplementary schedules** to the statement of cash flows include additional information reporting noncash investing and financing activities and cash paid for interest and income taxes.

- If the owners of a piece of land are willing to exchange it for shares of stock in a business, the balance sheet would, of course, change (with an increase in the asset portion of the balance sheet titled "Land" and a corresponding increase in the Owner's Equity portion of the balance sheet). The cash position of the business would not have changed because this would have been a **noncash transaction**.
- Any noncash investing and financing transactions undertaken by a company should be reported in a Supplementary Schedule of Noncash Investing and Finance Activities that is attached as a supplement to the SCF.
- Also included in the statement of cash flows (and required by the FASB) is the Supplementary Disclosure of Cash Flow Information, which reports cash paid during the year for interest and income taxes.
- A complete SCF should include:
  - A summary of cash inflows and outflows resulting from operating activities
  - A summary of cash inflows and outflows resulting from investing activities
  - A summary of cash inflows and outflows resulting from financing activities
  - Net changes in cash from the beginning to the ending of the accounting period
  - A supplementary schedule of noncash investing and financing activities (if applicable)
  - A supplementary disclosure of cash flow information
- For a complete illustration of building a Statement of Cash Flows for the Blue Lagoon Water Park Resort, see Figures 5.10, 5.11, 5.12, and *Go Figure!* in the text.

Statement of Cash Flows Analysis

- The SCF is not analyzed nearly as much as the income statement or balance sheet, partly because it is a relative newcomer to the managerial accounting world (it has only been required by the FASB since 1988).
- Many managers' jobs are more concerned with operations (income statement) and the effective use of assets (balance sheet) than cash management, which is often the job of the operations owner, **Chief Financial Officer (CFO),** controller, or accountant.
- One way to analyze a statement of cash flows is to first look at the sources and uses of funds identified by comparing last year's balance sheet with this year's balance sheet. By matching "like" dollar amounts of sources and uses of funds, you can surmise how funds were spent based on how funds were generated (see Figure 5.13).
- Another method of analyzing the statement of cash flows is to compare operating, investing, and financing activities from last year to this year (see Figure 5.14). The dollar or percentage variance shows changes from previously experienced levels, and will give you an indication of whether your numbers are improving, declining, or staying the same.

Go Figure!

To calculate the variance in investing activities, you would use the following formula:

> **Investing Activities This Year − Investing Activities Last Year = Variance**

The percentage variance for investing activities is determined as follows:

> **(Investing Activities This Year − Investing Activities Last Year)/Investing Activities Last Year = Percentage Variance**

An alternative and shorter formula for computing the percentage variance is as follows:

> **Variance/Investing Activities Last Year = Percentage Variance**

- While present cash flows in a business are certainly no absolute guarantee of future cash flows, it is a good indication of how well managers of the company are generating cash flows.
- For many investors and managers, a business's free cash flow is an important measure of its economic health. **Free cash flow** is simply the amount of cash a business generates from its operating activities *minus* the amount of cash it must spend on its investment activities and capital expenditures. Thus, free cash flow is considered a good measure of a company's ability to pay its debts, ensure its growth, and pay (if applicable) its investors in the form of dividends.
- See *Go Figure!* in the text for an illustration of free cash flow.
- A company with a positive free cash flow can grow and invest its excess cash in its own expansion or alternative investments. If a company has a negative free cash flow it will need to supplement its cash from other sources, such as borrowing funds or seeking additional investors.
- The formula for free cash flow is:

	Net cash provided from operating activities
Less	**Cash used to acquire property and equipment**
Equals	**Free Cash Flow**

Key Terms & Concepts Review
Match the key terms with their correct definitions.

1. Statement of changes in financial position   _____
   a. The result of all the transactions and events involving the purchase and sale of marketable securities, investments, land, buildings, equipment, and other assets not generally purchased for resale.

2. Statement of cash flows (SCF) _____
3. Consolidated income statement _____
4. Solvency _____
5. Cash _____
6. Cash equivalents _____
7. Sources and uses of funds _____
8. Condensed income statement _____
9. Statement of retained earnings _____
10. Cash flow from operating activities _____
11. Cash flow from investing activities _____
12. Cash flow from financing activities _____
13. Principal (loan) _____
14. Net changes in cash _____
15. Supplementary schedules _____

b. The result of all of the transactions and events that normally make up a business's day to day activities.

c. Corporate officer responsible for managing the financial operations of a company.

d. A statement that intends to indicate how cash inflows and outflows affect the business during a specific accounting period.

e. Non monetary exchange of the value of an investing or financing activity.

f. Inflows and outflows of money affecting the cash position.

g. A representation of all cash inflows minus cash outflows from operating, investing, and financing activities.

h. The amount of cash a business generates from its operating activities *minus* the amount of cash it must spend on its investment activities and capital expenditures.

i. A statement that reports the changes in the accumulated account of profits over the life of the business that have not been distributed as dividends from last year to this year.

j. Additional information on the statement of cash flows that reports noncash investing and financing activities and cash paid for interest and income taxes.

k. A statement that shows all sources and uses of funds from operating, investing, and financing activities of a business.

l. Short-term, temporary investments such as treasury bills, certificates of deposit, or commercial paper that can be quickly and easily converted to cash.

m. The ability of a business to pay its debts as they become due.

n. A statement that combines the revenue, expense, and profit information from each individual sub-statement of a company.

o. The money held in cash banks, checking or savings accounts, electronic fund transfers from payment card companies, and certificates of deposit.

16. Noncash transaction _____ p. A statement that reports the revenues, expenses, and profits in a summary format, absent of specific departmental and undistributed expense details.

17. Chief Financial Officer (CFO) _____ q. The amount borrowed on a loan.

18. Free cash flow _____ r. The result of all the transactions and events involving buying or selling company stocks, payment of dividends, and borrowing and repayment of short and long-term debt.

Discussion Questions

1. Explain the reason cash flows are so critically important to the operation of a successful business.

2. Describe Figure 5.5 in the text and explain how it is used to determine the sources and uses of funds.

3. List the five major components of the Statement of Cash Flows.

4. Describe two methods of analyzing the Statement of Cash Flows (as a whole).

Quiz Yourself

Choose the letter of the best answer to the questions listed below.

1. The report which indicates how cash inflows and outflows affected a business during a specific accounting period is called
   a. Cash accounting report
   b. Balance sheet
   c. Statement of cash flows
   d. Retained earnings statement

2. Solvency refers to a business's ability to
    a. Convert its assets to cash
    b. Pay its debts as they become due
    c. Pay out dividends for its owners
    d. Provide bonuses for its managers

3. When comparing assets from last year's balance sheet to this year's balance sheet, decreases in assets represent
    a. Sources of funds
    b. Uses of funds
    c. Increases in retained earnings
    d. Decreases in retained earnings

4. When comparing liabilities from last year's balance sheet to this year's balance sheet, decreases in liabilities represent
    a. Sources of funds
    b. Uses of funds
    c. Increases in fixed assets
    d. Decreases in fixed assets

5. Selling food and paying employees are examples of a restaurant's
    a. Investing activities
    b. Financing activities
    c. Supplementary activities
    d. Operating activities

6. Depreciation is a method of
    a. Marking down an asset that has been damaged
    b. Adjusting an asset's value to reflect its current market value
    c. Allocating the cost of a fixed asset over the useful life of the asset
    d. Allowing a company to pay for its fixed assets over a period of time

7. Selling marketable securities and buying a building are examples of
    a. Operating activities
    b. Investing activities
    c. Financing activities
    d. Supplementary activities

8. Borrowing and repaying borrowed money are examples of
    a. Financing activities
    b. Operating activities
    c. Supplementary activities
    d. Investing activities

9. The net investing activities this year at Peter's Pizza Parlor are $56,100 and the net investing activities last year were $48,300. Calculate the percentage variance.
   a. 13.9%
   b. 86.1%
   c. 16.2%
   d. 116.2%

10. The net cash flow from operating activities at Peter's Pizza Parlor is $98,750. Peter spent $10,800 on a new ice machine for his restaurant. Calculate Peter's free cash flow.
    a. $109,550
    b. $ 80,750
    c. $154,850
    d. $ 87,950

Chapter 6

Ratio Analysis

Highlights

* Purpose and Value of Ratios
* Types of Ratios
    * Liquidity Ratios
    * Solvency Ratios
    * Activity Ratios
    * Profitability Ratios
    * Investor Ratios
    * Hospitality Specific Ratios
* Comparative Analysis of Ratios
* Ratio Analysis Limitations
* Key Terms and Concepts Review
* Discussion Questions
* Quiz Yourself

Study Notes

Purpose and Value of Ratios

- Ratios are important in a variety of fields. This is especially true in the hospitality industry. If you are hospitality manager with a foodservice background, you already know about the importance of ratios. For examples of how ratios are used in foodservice, see *Go Figure!* in the text.
- A ratio is created when you divide one number by another. A special relationship (a **percentage**) results when the numerator (top number) used in your division is a *part* of the denominator (bottom number).
- In fraction form, a percentage is expressed as the part, or a portion of 100. Thus, 10 percent is written as 10 "over" 100 (10/100). In its common form, the "%" sign is used to express the percentage. If we say 10%, then we mean "10 out of each 100". The decimal form uses the (.) or decimal point to express the percent relationship. Thus, 10% is expressed as 0.10 in decimal form. The numbers to the right of the decimal point express the percentage.
- Each of these three methods of expressing percentages is used in the hospitality industry. To determine what percent one number is of another number, divide the number that is the part by the number that is the whole (see *Go Figure!* in the text).

- Many people become confused when converting from one form of percent to another. If that is a problem, remember the following conversion rules:
  - To convert from common form to decimal form, move the decimal two places to the left, that is, 50.00% = 0.50.
  - To convert from decimal form to common form, move the decimal two places to the right, that is, 0.50 = 50.00%.

Value of Ratios to Stakeholders

- All stakeholders who are affected by a business's profitability will care greatly about the effective operation of a hospitality business. These stakeholders may include:
  - Owners
  - Investors
  - Lenders
  - Creditors
  - Managers
- Each of these stakeholders may have different points of view of the relative value of each of the ratios calculated for a hospitality business. Owners and investors are primarily interested in their return on investment (ROI), while lenders and creditors are mostly concerned with their debt being repaid.
- At times these differing goals of stakeholders can be especially troublesome to managers who have to please their constituencies. One of the main reasons for this conflict lies within the concept of financial leverage.
- **Financial leverage** is most easily defined as the use of debt to be reinvested to generate a higher return on investment (ROI) than the cost of debt (interest). For an illustration of financial leverage, see *Go Figure!* in the text.
- Because of financial leverage, owners and investors generally like to see debt on a company's balance sheet because if it is reinvested well, it will provide more of a return on the money they have invested.
- Conversely, lenders and creditors generally do not like to see too much debt on a company's balance sheet because the more debt a company has, the less likely it will be able to generate enough money to pay off its debt.
- Ratios are most useful when they compare a company's actual performance to a previous time period, competitor company results, industry averages, or budgeted (planned for) results. When a ratio is compared to a standard or goal, the resulting differences (if differences exist) can tell you much about the financial performance (health) of the company you are evaluating.

Types of Ratios

- The most common way to classify ratios is by the information they provide the user. Managerial accountants working in the hospitality industry refer to:
  - Liquidity Ratios
  - Solvency Ratios

- o Activity Ratios
- o Profitability Ratios
- o Investor Ratios
- o Hospitality Specific Ratios
- Most numbers for these ratios can be found on a company's income statement, balance sheet, and statement of cash flows. (See Figures 6.1, 6.2, 6.3, and 6.4).
- Definitions, sources of data, formulas and examples of each ratio are summarized at the end of this section in the Ratio Summary Tables.
- Some managers use averages in the denominators of some ratios to smooth out excessive fluctuations from one period to the next. With the exception of inventory turnover, the ratios in this chapter will not use averages in the denominators.

Liquidity Ratios

- Liquidity is defined as the ease at which current assets can be converted to cash in a short period of time (less than 12 months). **Liquidity ratios** have been developed to assess just how readily current assets could be converted to cash, as well as how much current liabilities those current assets could pay.
- In this section we will examine three widely used liquidity ratios and working capital. These are:
  - o Current Ratio
  - o Quick (Acid-Test) Ratio
  - o Operating Cash Flows to Current Liabilities Ratio
  - o Working Capital

Current Ratio

- One of the most frequently computed liquidity ratios is the **current ratio.**
- **When current ratios are:**
  - o **Less than 1**: The business may have a difficult time paying its short term debt obligations because of a shortage of current assets.
  - o **Equal to 1**: The business has an equal amount of current assets and current liabilities.
  - o **Greater than 1**: The business has more current assets than current liabilities and should be in a good position to pay its bills as they come due.
- It might seem desirable for every hospitality business to have a high current ratio (because then the business could easily pay all of its current liabilities). That is not always the case.
  - o While potential creditors would certainly like to see a business in a position to readily pay all of its short-term debts, investors may be more interested in the financial leverage provided by short-term debts.
- The current ratio is so important to a hospitality business that lenders will frequently require that any business seeking a loan maintain a minimum current ratio during the life of any loan it is granted.

Quick (Acid-Test) Ratio

- Another extremely useful liquidity ratio is called the **quick ratio**. The quick ratio is also known as the **acid-test ratio**.
- The main difference between the current ratio formula and the quick ratio formula is the inclusion (or exclusion) of inventories and prepaid expenses. The purpose of the quick ratio is primarily to identify the relative value of a business's cash (and *quickly* convertible to cash) current assets.
- Investors and creditors view quick ratios in a manner similar to that of current ratios. Investors tend to prefer lower values for quick ratios, while creditors prefer higher ratios.

Operating Cash Flows to Current Liabilities Ratio

- The **operating cash flows to current liabilities ratio** relies on the operating cash flow portion of the overall statement of cash flows for its computation. It utilizes information from the balance sheet and the statement of cash flows.
- In general, investors and creditors view the operating cash flows to current liabilities ratio in a manner similar to that of the current and quick ratios.

Working Capital

- A measure that is related to the current and quick ratios is **working capital**. Although not a true ratio because it does not require that one number is divided by another number, it is a measure that many lenders require.
- Because of financial leverage, investors tend to prefer lower values for liquidity ratios, while creditors prefer higher values.

Solvency Ratios

- Just as liquidity ratios address the ability of a business to pay its short term debt, **solvency ratios** help managers evaluate a company's ability to pay long term debt. Solvency ratios are important because they provide lenders and owners information about a business's ability to withstand operating losses incurred by the business:
  - Solvency Ratio
  - Debt to Equity Ratio
  - Debt to Assets Ratio
  - Operating Cash Flows to Total Liabilities Ratio
  - Times Interest Earned Ratio

Solvency Ratio

- A business is considered **solvent** when its assets are greater than its liabilities. The **solvency ratio** compares a business's total assets to its total liabilities.

- This ratio is really a comparison between what a company "owns" (its assets) and what it "owes" those who do not own the company (liabilities).
- Creditors and lenders prefer to do business with companies that have a high solvency ratio (between 1.5 and 2.00) because it means these companies are likely to be able to repay their debts. Investors, on the other hand, generally prefer a lower solvency ratio, which may indicate that the company uses more debts as financial leverage.

Debt to Equity Ratio

- The **debt to equity ratio** is a measure used by managerial accountants to evaluate the relationship between investments that have been made by the business's lenders and investments that have been made by the business's owners.
- From a lender's perspective, the higher the lender's own investment (relative to the actual investment of the business's owners) the riskier is the investment.
- Owners seek to maximize their financial leverage and create total liabilities to total equity ratios in excess of 1.00.

Debt to Assets Ratio

- The **debt to assets ratio** compares a business's total liabilities to its total assets.
- As with the other solvency ratios, more debt will be favored by investors because of financial leverage and less debt will be favored by lenders to ensure repayment of loans.

Operating Cash Flows to Total Liabilities Ratio

- The **operating cash flows to total liabilities ratio** compares the cash generated by operating activities to the amount of total liabilities.
- In nearly all cases, both owners and lenders would like to see this ratio kept as high as possible because a high ratio indicates a strong ability to repay debt from the business's normal business operations

Times Interest Earned Ratio

- The **times interest earned ratio** compares interest expense to earnings before interest and taxes (EBIT). **Earnings before interest and taxes (EBIT)** are labeled as net operating income on the USALI.
- The higher this ratio, the greater the number of "times" the company could repay its interest expense with its earnings before interest and taxes.

Activity Ratios

- The purpose of computing **activity ratios** is to assess management's ability to effectively utilize the company's assets. Activity ratios measure the "activity" of a company's selected assets by creating ratios that measure the number of times

these assets *turn over* (are replaced), thus assessing management's *efficiency* in handling inventories and long-term assets. As a result, these ratios are also known as **turnover ratios** or **efficiency ratios.**
- In this section you will learn about the following activity ratios:
  o Inventory Turnover
  o Property and Equipment (Fixed Asset) Turnover
  o Total Asset Turnover

Inventory Turnover

- **Inventory turnover** refers to the number of times the total value of inventory has been purchased and replaced in an accounting period. In restaurants, we will calculate **food and beverage inventory turnover ratios** (refer to Figure 6.5)
- The obvious question is, "Are the food and beverage turnover ratios good or bad?" The answer to this question is relative to the **target** (desired) **turnover ratios.** For a discussion of food and beverage turnover ratio analysis, see *Go Figure!* in the text.
- A low turnover could occur because sales are less than expected, thus causing food to move slower out of inventory (bad). It could also mean that the food and beverage manager decided to buy more inventory each time (thus, making purchases fewer times) because of discount prices due to larger (bulk) purchases (good).
- A high turnover could occur because sales are higher than expected, thus causing food to move faster out of inventory (good). It could also mean that significant wastage, pilferage, and spoilage might have occurred causing food to move out of inventory faster, but *not* due to higher sales (bad).

Property and Equipment (Fixed Asset) Turnover

- The **property and equipment (fixed asset) turnover ratio** is concerned with fixed asset usage. **Fixed assets** consist of the property, building(s) and equipment actually owned by the business.
- A simple example will explain how to interpret this ratio. Assume that there is a fryer in the kitchen which generates $50,000 worth of revenue per year. A new fryer is purchased that generates revenues of $100,000 per year. The new fryer would have a fixed asset turnover ratio two times higher than that of the old fryer. The new fryer is more effective at generating revenues than the old fryer.
- The term "net" in any calculation generally means that something has been subtracted. When calculating the net property and equipment turnover ratio, "net" refers to the subtraction of accumulated depreciation.
- Creditors, owners, and managers like to see this ratio as high as possible because it measures how effectively net fixed assets are used to generate revenue.

Total Asset Turnover

- The **total asset turnover ratio** is concerned with total asset usage. Total assets consist of the current and fixed assets owned by the business.
- Restaurants may have higher ratios than hotels because hotels typically have more fixed assets (thus making the denominator larger and the ratio smaller).
- Creditors, owners, and managers like to see this ratio as high as possible because it measures how effectively total assets are used to generate revenue.

Profitability Ratios

- It is the job of management to generate profits for the company's owners, and **profitability ratios** measure how well management has accomplished this task.
- Profits must also be evaluated in terms of the size of investment in the business that has been made by the company's owners.
- There are a variety of profitability ratios used by managerial accountants:
  - Profit Margin
  - Gross Operating Profit Margin (Operating Efficiency)
  - Return on Assets
  - Return on Owner's Equity

Profit Margin

- **Profit margin** is the term managerial accountants use to describe the ability of management to provide a profit for the company's owners. This ratio compares the amount of net income generated by a business to the revenue it generated in the same time period.

Gross Operating Profit Margin

- The **gross operating profit margin** is also known as the **operating efficiency ratio.** In the opinion of most managerial accountants, it is a better measure of actual management effectiveness than is profit margin.

Return on Assets

- The original investment and profits paid back to owners are called **returns**. When developing profitability ratios, managerial accountants want to examine the size of an investor's return. **Return on assets (ROA)** is one such ratio.
- The importance of this ratio is easy to understand if you analyze it using a comparison of two companies (see *Go Figure!* in the text).
- Managerial accountants carefully review the ROA achieved in a business and compare it to industry averages, the business owner's own investment goals, and other valid benchmarks.

Return on Owner's Equity

- **Return on equity (ROE)** is a ratio developed to evaluate the rate of return on the personal funds actually invested by the owners (and/or shareholders) of the business.
- Even for individual business owners, the greatest interest is in determining the total amount they have personally invested and the total amount (after taxes) that investment will actually **yield** (return back to them).

Investor Ratios

- **Investor ratios** assess the performance of earnings and stocks of a company. Investors use these ratios to choose new stocks to buy and to monitor stocks they already own.
- Investors are interested in two types of returns from their stock investments: money that can be earned from the sale of stocks at higher prices than originally paid, and money that can be earned through the distribution of dividends.
- Investors use many different ratios to make decisions on investments:
  - Earnings per Share
  - Price/Earnings Ratio
  - Dividend Payout Ratio
  - Dividend Yield Ratio

Earnings Per Share

- **Earnings per share (EPS)** is of most interest to those who buy and sell stocks in publicly traded companies, and to those managers who operate these companies.
- This ratio is strongly affected by both the amount of money a company earns and the number of shares of stock its board of directors elects to issue to the public.

Price/Earnings Ratio

- For investors, the question of a stock's "value" is assessed, in part, by computing its **price/earnings (P/E) ratio**. The price/earnings ratio compares market price per share to earnings per share.
- Neither the numerator nor denominator of this ratio is located directly on the company's financial statements, although the calculation for earnings per share does use information from the income statement.
- Price/earnings ratios for hospitality industry stocks vary greatly. The question of whether a specific P/E ratio is a "good" one or a "bad" one is dependent upon the goals of the investor as well as that specific investor's view of the company's profitability and growth potential.

Dividend Payout Ratio

- The **dividend payout ratio** compares dividends to be paid to stockholders with earnings per share. This ratio is determined annually by the company's board of directors based on the desired amount to pay stockholders (dividends) and the amount to be reinvested (retained) in the company.

Dividend Yield

- **Dividend yield ratio** compares the dividends per share to the market price per share.
- The dividend yield ratio can be used by investors who wish to find stocks that will supplement their income with dividends.

Hospitality Specific Ratios

- The numbers used to create these ratios are often found on daily, weekly, monthly or yearly operating reports that managers design to fit their operational needs.

Hotel Ratios

- The hotel-specific ratios in this section are:
    - Occupancy Percentage
    - Average Daily Rate (ADR)
    - Revenue Per Available Room (RevPAR)
    - Revenue Per Available Customer (RevPAC)
    - Cost Per Occupied Room (CPOR)

Occupancy Percentage

- Hotel managers and owners are interested in the **occupancy percentage** (percentage of rooms sold in relation to rooms available for sale) because occupancy percentage is one measure of a hotel's effectiveness in selling rooms.
- Most hotels will have rooms, floors, or entire wings at various times of the year that are **out of order (OOO)**, meaning that repairs, renovation, or construction is being done and the rooms are not sellable. When calculating occupancy percentage, adjustments (subtractions) should be made to the denominator to account for rooms that are out of order.
- Complimentary occupancy (percentage of rooms provided on a **complimentary (comp)** basis or free of charge), average occupancy per room (average number of guests occupying each room), and multiple occupancy (percentage of rooms occupied by two or more people) are all variations of room occupancy.
- Occupancy percentage can used to compare a hotel's performance to previous accounting periods, to forecasted or budgeted results, to similar hotels, and to published industry averages or standards.

- Industry averages and other hotel statistics are readily available through companies such as **Smith Travel Research (STR)**. Smith Travel Research is a compiler and distributor of hotel industry data.

Average Daily Rate (ADR)

- Hoteliers are interested in the **average daily rate (ADR)** they achieve during an accounting period. ADR is the average amount for which a hotel sells its rooms.
- Most hotels offer their guests the choice of several different **room types**. Each specific room type will likely sell at a different nightly rate. When a hotel reports its total nightly revenue, however, its overall *average* daily rate is computed.

Revenue Per Available Room (RevPAR)

- High occupancy percentages can be achieved by selling rooms inexpensively, and high ADRs can be achieved at the sacrifice of significantly lowered occupancy percentages. Hoteliers have developed a measure of performance that *combines* these two ratios to compute **revenue per available room (RevPAR)**.
- For an illustration of how RevPAR can be used to compare two hotels, see *Go Figure!* in the text.

Revenue Per Available Customer (RevPAC)

- Hotel managers are interested in the **revenue per available customer (RevPAC)** (revenues generated by each customer) because guests spend money on many products in a hotel in addition to rooms.
- RevPAC is especially helpful when comparing two groups of guests. Groups that generate a high RevPAC are preferable to groups that generate a lower RevPAC.

Cost Per Occupied Room (CPOR)

- **Cost per occupied room (CPOR)** is a ratio that compares specific costs in relation to number of occupied rooms. CPOR is computed for guest amenity costs, housekeeping costs, laundry costs, in-room entertainment costs, security costs, and a variety of other costs.
- CPOR can be used to compare one type of cost in a hotel to other hotels within a chain, a company, a region of the country, or to any other standard deemed appropriate by the hotel's managers or owners.

Restaurant Ratios

- The restaurant-specific ratios in this section are:
    - Cost of Food Sold (Cost of Sales: Food)
    - Cost of Beverage Sold (Cost of Sales: Beverage)
    - Food Cost Percentage
    - Beverage Cost Percentage
    - Average Sales per Guest (Check Average)

- Seat Turnover

Cost of Food Sold (Cost of Sales: Food)

- **Cost of food sold (cost of sales: food)** is the dollar amount of all food expenses incurred during the accounting period. **Cost of *goods* sold** is a general term for cost of any products sold. For restaurants, cost of goods sold as referenced in the inventory turnover section of this chapter refers to cost of food sold and cost of beverage sold.
- The formula for cost of food sold follows:

> **Beginning Inventory**
> **+ Purchases**
> **= Food Available for Sale**
> **- Ending Inventory**
> **= Cost of Food Consumed**
> **- Employee Meals**
> **= Cost of Food Sold**

- **Beginning inventory** is the dollar value of all food on hand at the beginning of the accounting period. This inventory may be referred to by its synonymous term, **opening inventory**. Beginning inventory is determined by completing an actual count and valuation of the products on hand.
- **Purchases**, as used in this formula, are the sum costs of all food purchased during the accounting period. **Food available for sale** is the sum of the beginning inventory plus the value of all purchases.
- **Ending inventory** refers to the dollar value of all food on hand at the end of the accounting period. This inventory may be referred to by its synonym, **closing inventory**. It also is determined by completing a physical inventory.
- The **cost of food consumed** is the actual dollar value of all food used, or consumed, by the operation. This is not merely the value of all food sold, but rather the value of all food no longer in the establishment. This includes the value of any meals eaten by employees and also food that is lost due to wastage, pilferage, and spoilage.
- **Cost of *goods* consumed** is a general term for cost of any products consumed. For restaurants, cost of goods consumed refers to cost of food consumed and cost of beverage consumed.
- **Employee meals** are a labor-related expense. The cost of this benefit, if provided, should be accounted for in the Employee Benefits line item of the Operating Expenses section of the income statement. Since this expense belongs under Employee Benefits, it is subtracted from cost of food consumed to yield the cost of food sold (cost of sales) on the income statement.
- **Transfers out** are items that have been transferred out of one unit to another, and **transfers in** are items that have been transferred in to one unit from another.
- The formula for cost of food sold in this situation would be as follows:

> **Beginning Inventory**
> **+ Purchases**
> **= Food Available for Sale**
> **- Ending Inventory**
> **= Cost of Food Consumed**
> **- Value of Transfers Out**
> **+ Value of Transfers In**
> **- Employee Meals**
> **= Cost of Food Sold**

- See *Go Figure!* in the text for an example of the computation of Cost of Food Sold (Cost of Sales: Food).

Cost of Beverage Sold (Cost of Sales: Beverage)

- **Cost of beverage sold (cost of sales: beverage)** is the dollar amount of all beverage expenses incurred during the accounting period. The cost of beverage sold is calculated in the same way as cost of food sold except that the products are alcoholic beverages (beer, wine, and spirits). Employee meals are not subtracted because employees are not drinking alcoholic beverages.
- The computation of cost of beverage sold is as follows:

> **Beginning Inventory**
> **+ Purchases**
> **= Beverage Available for Sale**
> **- Ending Inventory**
> **= Cost of Beverage Sold**

- The cost of beverage sold formula is amended for transfers as follows:

> **Beginning Inventory**
> **+ Purchases**
> **= Beverage Available for Sale**
> - Ending Inventory
> - Value of Transfers Out
> **+ Value of Transfers In**
> **= Cost of Beverage Sold**

- See *Go Figure!* in the text for an example of the computation of Cost of Beverage Sold (Cost of Sales: Beverage).
- Accurate beginning and ending inventory figures must be maintained if an operation's true food and beverage cost data are to be computed. Ending inventory for one accounting period becomes beginning inventory for the next period.

Food Cost Percentage

- A restaurant's food cost percentage is the ratio of the restaurant's cost of food sold (cost of sales: food) and its food revenue (sales). Thus, **food cost percentage** represents the portion of food sales that was spent on food expenses.
- The calculation for food cost percentage is:

Beverage Cost Percentage

- A restaurant's beverage cost percentage is the ratio of the restaurant's cost of beverage sold (cost of sales: beverage) and its beverage revenue (sales). Thus, **beverage cost percentage** represents the portion of beverage sales that was spent on beverage expenses.
- For most restaurants, the beverage cost percentage, its management, and its control are extremely important because alcoholic beverages represent an expensive cost and a serious security problem.

Labor Cost Percentage

- Restaurateurs are very interested in the **labor cost percentage**, which is the portion of total sales that was spent on labor expenses. Labor costs include salaries and wages and other labor-related expenses such as employee benefits.
- Increasing total sales (the denominator in the formula) will help decrease the labor cost percentage even as total dollars spent on labor increases, assuming some of the labor costs represent fixed expenses such as salaries. It is typically *not* in the best interest of restaurant operators to reduce the total amount they spend on labor. In most foodservice situations, managers *want* to serve more guests, and that typically requires additional staff.
- The labor cost percentage is important because it helps managers relate the amount of products they sell to the cost of the staff needed to sell them. Many managers feel it is more important to control labor costs than product costs because, for many of them, labor and labor-related costs comprise a larger portion of their operating budgets than do the food and beverage products they sell.

Average Sales per Guest (Check Average)

- **Average sales per guest** is the average amount of money spent per customer during a given accounting period. Average sales per guest is also commonly known as **check average**.
- Most **point of sale (POS) systems** (computer systems used for tracking sales data) will tell you the amount of revenue you have generated in a selected time period, the number of guests you have served, and the average sales per guest.
- This measure of "sales per guest" is important because it carries information needed to monitor menu item popularity, estimate staffing requirements, and even determine purchasing procedures. It also allows a financial analyst to measure a chain's effectiveness in increasing sales to its current guests, rather than increasing sales simply by opening additional restaurants.

- The check average ratio can be used to compare a restaurant's performance to previous accounting periods, to forecasted or budgeted results, to similar restaurants, and to published industry averages or standards. Industry averages and other restaurant statistics are readily available through publications such as the ***Restaurant Industry Operations Report*** published by the National Restaurant Association.

Seat Turnover

- To evaluate a restaurant's effectiveness in "turning tables" the seat turnover ratio is a popular one. **Seat turnover** measures the number of times seats change from the current diner to the next diner in a given accounting period.
- This ratio does not use information from the income statement, the balance sheet, or the statement of cash flows. It is one of the many financial ratios used by managers in the hospitality industry for which the data is generated completely internally.
- The interpretation of this ratio must be carefully undertaken because its value is so greatly determined by the "covers served" that comprises this ratio's numerator. Often, management must make decisions about the definition of a cover, such as:
    - Is a guest who accompanies another diner but does not eat considered a cover?
    - Must a cover purchase a minimum number of menu items or dollar value of items to be considered a cover?
    - Should the definition of a cover change from breakfast, lunch, and dinner?
- Managers want to see this ratio as high as reasonably possible because the seat turnover ratio is an important indication of a restaurant's ability to effectively utilize its "seats" to sell it products.
- See Figure 6.7 for a summary of the ratios used by managers.

Comparative Analysis of Ratios

- Like many other types of financial data, a company's financial ratios are often compared to previous accounting periods, to forecasted or budgeted results, or to published industry averages or standards (see Figure 6.8 and 6.9).

Ratio Analysis Limitations

- One weakness inherent in an over-dependency on financial ratios is that ratios, by themselves, may be less meaningful unless compared to those of previous accounting periods, budgeted results, industry averages, or similar properties.
- Another limitation is that financial ratios do not measure a company's **intellectual capital assets** such as brand name, potential for growth, and intellectual or human capital when assessing a company's true worth. See Figure 6.10 for a list of some of the intellectual capital assets that should be analyzed in addition to financial ratios to assess the health and worth of a company.

Liquidity Ratio Summary Table

Ratio Name	Definition	Source of Data	Formula
Current Ratio	Current ratio shows the firm's ability to cover its current liabilities with its current assets.	Numerator: Balance Sheet Denominator: Balance Sheet	$\dfrac{\text{Current Assets}}{\text{Current Liabilities}}$
Quick (Acid-Test) Ratio	Quick ratio shows the firm's ability to cover its current liabilities with its *most liquid* current assets.	Numerator: Balance Sheet Denominator: Balance Sheet	$\dfrac{\text{Cash + marketable securities + accounts receivable}}{\text{Current liabilities}}$ *or* $\dfrac{\text{Current assets} - (\text{inventories + prepaid expenses})}{\text{Current liabilities}}$
Operating Cash Flows to Current Liabilities Ratio	Operating cash flows to current liabilities ratio shows the firm's ability to cover its current liabilities with its operating cash flows.	Numerator: Statement of cash flows Denominator: Balance sheet	$\dfrac{\text{Operating cash flows}}{\text{Current liabilities}}$
Working Capital	Working capital is the difference between current assets and current liabilities.	Numerator: Balance Sheet Denominator: Balance Sheet	Current assets − Current liabilities

Solvency Ratio Summary Table

Ratio Name	Definition	Source of Data	Formula
Solvency Ratio	Solvency ratio shows the firms ability to cover its total liabilities with its total assets.	Numerator: Balance Sheet Denominator: Balance Sheet	$\dfrac{\text{Total assets}}{\text{Total liabilities}}$

Ratio Name	Definition	Source of Data	Formula
Debt to Equity Ratio	Debt to equity ratio compares total liabilities to owners' equity.	Numerator: Balance Sheet Denominator: Balance Sheet	$\dfrac{\text{Total liabilities}}{\text{Total owner's equity}}$
Debt to Assets Ratio	Debt to assets ratio shows the percentage of assets financed through debt.	Numerator: Balance Sheet Denominator: Balance Sheet	$\dfrac{\text{Total liabilities}}{\text{Total assets}}$
Operating Cash Flows to Total Liabilities Ratio	Operating cash flows to total liabilities ratio shows the firm's ability to cover its total liabilities with its operating cash flows.	Numerator: Statement of cash flows Denominator: Balance sheet	$\dfrac{\text{Operating cash flows}}{\text{Total liabilities}}$
Times Interest Earned Ratio	Times interest earned shows the firm's ability to cover interest expenses with earnings before interest and taxes.	Numerator: Income statement Denominator: Income statement	$\dfrac{\text{Earnings Before Interest and Taxes (EBIT)}}{\text{Interest Expense}}$

Activity Ratio Summary Table

Ratio Name	Definition	Source of Data	Formula
Food Inventory Turnover Ratio	Food inventory turnover shows the speed (# of times) that food inventory is replaced (turned) during a year	Numerator: Income statement Denominator: Balance sheet	$\dfrac{\text{Cost of food consumed}}{\text{Average food inventory*}}$ *(Beginning food inventory + ending food inventory)/2

Ratio Name	Definition	Source of Data	Formula
Beverage Inventory Turnover Ratio	Beverage inventory turnover shows the speed (# of times) that beverage inventory is replaced (turned) during a year	Numerator: Income statement Denominator: Balance sheet	$\dfrac{\text{Cost of beverage consumed}}{\text{Average beverage inventory*}}$ *(Beginning beverage inventory + ending beverage inventory)/2
Property and Equipment (Fixed Asset) Turnover Ratio	Property and equipment turnover ratio shows management's ability to effectively use net property and equipment to generate revenues.	Numerator: Income statement Denominator: Balance sheet	$\dfrac{\text{Total Revenue}}{\text{Net Property and Equipment}}$
Total Asset Turnover Ratio	Total asset turnover shows management's ability to effectively use total assets to generate revenues.	Numerator: Income statement Denominator: Balance sheet	$\dfrac{\text{Total Revenue}}{\text{Total Assets}}$

Profitability Ratio Summary Table

Ratio Name	Definition	Source of Data	Formula
Profit Margin	Profit margin shows management's ability to generate sales, control expenses, and provide a profit.	Numerator: Income statement Denominator: Income statement	$\dfrac{\text{Net income}}{\text{Total revenue}}$
Gross Operating Profit Margin (Operating Efficiency Ratio)	Gross operating profit margin shows management's ability to generate sales, control expenses, and provide a gross operating profit.	Numerator: Income statement Denominator: Income statement	$\dfrac{\text{Gross operating profit}}{\text{Total revenue}}$

Ratio Name	Definition	Source of Data	Formula
Return on Assets Ratio	Return on assets shows the firm's ability to use total assets to generate net income.	Numerator: Income statement Denominator: Balance sheet	$\dfrac{\text{Net income}}{\text{Total assets}}$
Return on Equity Ratio	Return on equity shows the firm's ability to use owner's equity to generate net income.	Numerator: Income statement Denominator: Balance sheet	$\dfrac{\text{Net income}}{\text{Total owners' equity}}$
Earnings Per Share Ratio	Earnings per share compares net income to common shares.	Numerator: Income statement Denominator: Statement of Retained Earnings and Investor Information	$\dfrac{\text{Net income}}{\text{Total number of common shares outstanding}}$
Price/Earnings (P/E) Ratio	Price/earnings ratio shows the perception of the firm in the market about future earnings growth of the company.	Numerator: Statement of Retained Earnings and Investor Information Denominator: Statement of Retained Earnings and Investor Information	$\dfrac{\text{Market price per share}}{\text{Earnings per share}}$
Dividend Payout Ratio	Dividend payout ratio shows the percentage of net income that is to be paid out in dividends.	Numerator: Statement of Retained Earnings and Investor Information Denominator: Statement of Retained Earnings and Investor Information	$\dfrac{\text{Dividend per share}}{\text{Earnings per share}}$
Dividend Yield Ratio	Dividend yield shows the stockholders' return on investment paid in dividends.	Numerator: Statement of Retained Earnings and Investor Information Denominator: Statement of Retained Earnings and Investor Information	$\dfrac{\text{Dividend per share}}{\text{Market price per share}}$

Hospitality Ratio Summary Table

Ratio Name	Definition	Source of Data	Formula
Occupancy Percentage	Occupancy % shows percentage of rooms sold in relation to rooms available for sale	Numerator: Operating Reports Denominator: Operating Reports	$\dfrac{\text{Rooms Sold}}{\text{Rooms Available for Sale}}$
Average Daily Rate (ADR)	Average daily rate shows average amount for which a hotel sells its rooms	Numerator: Operating Reports Denominator: Operating Reports	$\dfrac{\text{Total Rooms Revenue}}{\text{Total Number of Rooms Sold}}$
Revenue per Available room (RevPAR)	RevPar shows revenues generated by each available room	Numerator: Operating Reports Denominator: Operating Reports	Occupancy % x ADR *or* $\dfrac{\text{Total Rooms Revenue}}{\text{Rooms Available for Sale}}$
Cost per Occupied Room (CPOR)	Cost per occupied room compares specific costs in relation to number of occupied rooms	Numerator: Operating Reports Denominator: Operating Reports	$\dfrac{\text{Cost Under Examination}}{\text{Rooms Occupied}}$
Food Cost Percentage	Food Cost Percentage represents portion of food sales spent on food expenses	Numerator: Operating Reports Denominator: Operating Reports	$\dfrac{\text{Cost of Food Sold}}{\text{Food Sales}}$
Beverage Cost Percentage	Beverage Cost Percentage represents portion of Beverage sales spent on Beverage expenses	Numerator: Operating Reports Denominator: Operating Reports	$\dfrac{\text{Cost of Beverage Sold}}{\text{Beverage Sales}}$
Labor Cost Percentage	Labor Cost Percentage represents portion of total sales spent on labor expenses	Numerator: Operating Reports Denominator: Operating Reports	$\dfrac{\text{Cost of Labor*}}{\text{Total Sales}}$ * Cost of labor = salaries + wages + employee benefits

Ratio Name	Definition	Source of Data	Formula
Average Sales Per Guest (Check Average)	Average sales per guest is average amount of money spent per customer during given accounting period	Numerator: Operating Reports Denominator: Operating Reports	$\dfrac{\text{Total Sales}}{\text{Number of Guests Served}}$
Seat Turnover	Seat turnover shows number of times seats change from current diner to another diner in given accounting period	Numerator: Operating Reports Denominator: Operating Reports	$\dfrac{\text{Covers Served}}{\text{Number of Seats} \times \text{Number of Operating Days in Period}}$

Key Terms & Concepts Review

Match the key terms with their correct definitions.

1. Percentage _____
2. Financial leverage _____
3. Liquidity ratios _____
4. Current ratio _____
5. Quick (acid test) ratio _____
6. Operating cash flows to current liabilities ratio _____
7. Working capital _____
8. Solvency ratios (as a group of ratios) _____
9. Solvent _____
10. Solvency ratio (as a single ratio) _____
11. Debt to equity ratio _____
12. Debt to assets ratio _____
13. Operating cash flows to total liabilities ratio _____
14. Times interest earned ratio _____
15. Earnings before interest and taxes (EBIT) _____
16. Activity ratios _____

a. Ratio that shows management's ability to effectively use total assets to generate revenues.
b. Ratio that compares total liabilities to owners' equity.
c. Also referred to as gross operating profit margin.
d. The price, on average, for which a hotel sells its rooms.
e. Ratio that shows the percentage of net income that is to be paid out in dividends.
f. A company that collects and distributes summaries of hotel financial and operational data related to historical performance and prepares industry averages and other hotel statistics that are readily available through companies.
g. Return on investment.
h. Rooms that usually sell at different nightly rates and may include standard sized rooms, upgraded rooms such as parlor or whirlpool suites, connecting rooms, or very large suites.
i. A measure of performance that identifies the amount of revenue generated by each customer.
j. Ratio that shows management's ability to generate sales, control expenses, and provide a gross operating profit.
k. The state of a business when its assets are greater than its liabilities.
l. Ratio that shows the firm's ability to cover its total liabilities with its operating cash flows.
m. Ratio that shows the firm's ability to cover its current liabilities with its current assets.
n. The use of debt to be reinvested to generate a higher return on investment than the cost of debt (interest).
o. Ratio that shows the perception of the firm in the market about future earnings growth of the company.
p. The percentage of net income to revenues.

17. Turnover ratios _____
18. Efficiency ratios _____
19. Inventory turnover _____
20. Food inventory turnover _____
21. Beverage inventory turnover _____
22. Target turnover ratios _____
23. Property and equipment (fixed asset) turnover ratio _____
24. Fixed assets _____
25. Total asset turnover ratio _____
26. Profitability ratios _____
27. Profit margin _____
28. Gross operating profit margin _____
29. Operating efficiency ratio _____
30. Returns _____
31. Return on assets (ROA) _____
32. Return on equity (ROE) _____

q. Ratio that shows the percentage of rooms sold in relation to rooms available for sale.
r. The original investment and profits paid back to owners.
s. Term for a room that is provided free of charge.
t. Ratio that shows the firm's ability to cover interest expenses with earnings before interest and taxes.
u. Group of ratios that assesses the firm's ability to cover its total liabilities with its total assets
v. Ratio that shows the firm's ability to use owners' equity to generate net income.
w. Ratios that measure the number of times assets are replaced, thus assessing management's efficiency in handling inventories and long-term assets.
x. The difference between current assets and current liabilities.
y. Ratio that shows the number of times the total value of inventory has been purchased and replaced in an accounting period.
z. Ratio that shows stockholders' return on investment paid in dividends.
aa. The net operating income on the USALI.
bb. Ratio that compares total liabilities to total assets.
cc. A measure of performance that combines occupancy percentage and average daily rate.
dd. Ratio that shows the firm's ability to cover its current liabilities with its operating cash flows.
ee. Group of ratios that assesses how readily current assets could be converted to cash, as well as how much current liabilities those current assets could pay.
ff. Term for a room that is not sellable because repairs, renovation, or construction is being done.

33. Yield (investment) _____
34. Investor ratios _____
35. Earnings per share (EPS) _____
36. Price/earnings (PE) ratio _____
37. Dividend payout ratio _____
38. Dividend yield ratio _____
39. Occupancy percentage _____
40. Out of order (OOO) _____
41. Complimentary (comp) _____
42. Smith Travel Research (STR) _____
43. Star Reports _____
44. Average daily rate (ADR) _____
45. Room types _____
46. Revenue per available room (RevPAR) _____
47. Revenue per available customer (RevPAC) _____
48. Cost per occupied room (CPOR) _____

gg. Statistical reports compiled by Smith Travel Research.
hh. Ratio that shows the speed (# of times) that food inventory is replaced during a year.
ii. Ratio that shows the firm's ability to cover its current liabilities with its *most liquid* current assets.
jj. Ratio that shows the speed (# of times) that beverage inventory is replaced during a year.
kk. Also referred to as Turnover ratios.
ll. Group of ratios that shows management's ability to effectively utilize the company's assets.
mm. A relationship between two numbers in which the numerator (top number) is divided by the denominator (bottom number).
nn. Group of ratios that shows the performance of earnings and stocks of a company.
oo. Group of ratios that measures of how effectively management has generated profits for a company's owners.
pp. Ratio that shows management's ability to effectively use net property and equipment to generate revenues.
qq. Ratio that shows the desired turnover rate for inventory.
rr. Ratio that shows the firm's ability to use total assets to generate net income.
ss. Ratio that compares specific costs in relation to number of occupied rooms.
tt. Ratio that shows the comparison of a business's total assets to its total liabilities.
uu. Assets which management intends to keep for a period longer than one year including the property, building(s), and equipment owned by a business.
vv. Ratio that compares net income to common shares.

Discussion Questions

1. Use three methods of expressing percentages to express the value of 10 percent.

2. List the six types of ratios and describe how they are used by managerial accountants.

3. List five industry-specific ratios used by hotels.

4. List seven industry-specific ratios used by restaurants.

Quiz Yourself
Choose the letter of the best answer to the questions listed below.

1. Ratios which assess the ease at which current assets can be converted to cash and how much current liabilities those current assets could pay are called
    a. Solvency ratios
    b. Profitability ratios
    c. Investor ratios
    d. Liquidity ratios

2. Ratios which evaluate a business's ability to pay long-term debt are called
    a. Profitability ratios
    b. Activity ratios
    c. Solvency ratios
    d. Liquidity ratios

3. Ratios which assess management's ability to effectively utilize the company's assets are called
    a. Activity ratios
    b. Profitability ratios
    c. Liquidity ratios
    d. Solvency ratios

4. Ratios which measure how well management has generated profits for the company's owners are called
   a. Solvency ratios
   b. Profitability ratios
   c. Activity ratios
   d. Investor ratios

5. Ratios which assess the performance of earnings and stocks of a company are called
   a. Profitability ratios
   b. Solvency ratios
   c. Investor ratios
   d. Activity ratios

6. Calculate the occupancy percentage for a hotel when there are 950 rooms available for sale and 685 rooms are sold on a given night, with revenues of $84,940.
   a. 89.4%
   b. 55.4%
   c. 13.1%
   d. 72.1%

7. Calculate the ADR for a hotel where there are 950 rooms available for sale and 685 rooms are sold on a given night, with revenues of $84,940.
   a. $ 89.41
   b. $124.00
   c. $232.71
   d. $320.52

Questions 8, 9 & 10 are based on the following information from Ruth's Ribeye Steak House operating records for last year:

Total Sales	$1,368,211
Total Food Sales	$1,023,426
Cost of Food Sold	$ 335,423
Cost of Labor	$469,075
Number of Guests Served	31,003

8. Calculate the food cost percentage for Ruth's Ribeye Steak House.
   a. 32.8%
   b. 42.7%
   c. 70.7%
   d. 44.1%

9. Calculate the labor cost percentage for Ruth's Ribeye Steak House.
    a. 45.8%
    b. 34.3%
    c. 15.1%
    d. 29.2%

10. Calculate the check average for Ruth's Ribeye Steak House.
    a. $33.01
    b. $64.37
    c. $44.13
    d. $86.05

Part III: Management of Revenue and Expense

Chapter 7

Food and Beverage Pricing

Highlights

* Factors Affecting Menu Pricing
* Assigning Menu Prices
* Menu Price Analysis
* Matrix Analysis
* Goal Value Analysis
* Key Terms and Concepts Review
* Discussion Questions
* Quiz Yourself

Study Notes

Factors Affecting Menu Pricing

- Perhaps no area of hospitality management is less well understood than the area of pricing food and beverage products. Some of the most common factors affecting menu prices include one or more of the following:

Factors Influencing Menu Price
1. Local Competition
2. Service Levels
3. Guest Type
4. Product Quality
5. Portion Size
6. Ambience
7. Meal Period
8. Location
9. Sales Mix

- **Local Competition.** The price a competitor charges for his or her product can be useful information in helping you arrive at your own selling price.
- **Service Levels.** As the personal level of service increases, costs increase and thus prices must also be increased.

- **Guest Type**. A thorough analysis of who your guests are and what they value most is critical to the success of any restaurant's pricing strategy.
- **Product Quality**. You should select the quality level that best represents your guests' anticipated desires as well as your own operational goals, and then price your products accordingly.
- **Portion Size**. The effect of portion size on menu price is significant. It will be your job to establish and maintain strict control over portion size.
- **Ambiance**. Prices may be somewhat higher if the quality of products and ambiance also support the price structure.
- **Meal Period**. In some cases, diners expect to pay more for an item served in the evening than for that same item served at a lunch period.
- **Location**. A location can be a good for business or bad for business. If it is good, menu prices may reflect that fact. If a location is indeed bad, menu prices may need to be lower.
- **Sales Mix**. Sales mix has the most influence on a manager's menu pricing decisions. **Sales mix** refers to the frequency with which specific menu items are selected by guests.

Assigning Menu Prices

- There should be a clear and direct relationship between a restaurant's profits and its menu prices. Menu item pricing is related to revenue, costs (expenses), and profits by virtue of the following basic formula that you learned in Chapter 1:

> **Revenue - Expense = Profit**

- It is important to understand that revenue and price are not synonymous terms. Revenue refers to the amount spent by *all* guests, while price refers to the amount charged to *one* guest. Thus, total revenue is generated by the following formula:

> **Price x Number Sold = Total Revenue**

- The economic laws of **supply and demand** state that, for most products purchased by consumers, as the price of an item increases, the number of those items sold will generally decrease. Conversely, as the price of an item decreases, the number of those items sold will generally increase. For this reason, price increases must be evaluated based on their impact on total revenue and not on price alone.
- To illustrate the relationship of pricing to total revenue, refer to Figure 7.1, which illustrates the possible effects of this price increase on total revenue in a single unit. Increasing prices without giving added value can result in higher prices but, frequently, lower revenues because of reduced guest counts.
- Guests demand a good price/value relationship when making a purchase. The **price/value relationship** reflects guests' view of how much value they are receiving for the price they are paying.

Go Figure!

Assume that Sofia wants to raise the price of cheesecake sold in her downtown delicatessen. If she raises the price by 50 cents, how can she calculate the number that must be sold to keep her revenue constant? Sofia can use the following computations.

Current Number Sold X Current Price = Current Revenue

To calculate the number of newly priced cheesecakes that must be sold to maintain Sofia's current revenue she makes the following computation:

$$\frac{\text{Current Revenue}}{\text{New Price}} = \text{Number That Must Be Sold}$$

The actual results she achieves will depend largely on the price/value relationship her customers perceive regarding the higher priced cheesecake.

Marketing Approaches to Pricing

- The prices of the items sold on a menu can represent a variety of concepts. For example, when Ruth Chris, the famous New Orleans steakhouse restaurant group, sets the price for a steak on its menu, it seeks to tell its customers, "Come here for quality!" When Wendy's selects items for its 99 cent menu, it seeks to tell customers "Come here for value!"
- In a sales approach to marketing, the goal is to *maximize volume* (number of covers sold). Increased customer counts should result in maximized total operational revenues. This approach works best when service levels are limited, the products sold are easily produced, and the cost of providing the product can reliably and consistently be controlled.
- Other managers, usually in full-service restaurants, use the marketing philosophy of *maintaining your current competitive position* relative to the other restaurants in your market that target the same customers as you. Restaurateurs utilizing this approach feel that guests are primarily price conscious and will not pay "more" for the menu items at their restaurants than they would pay at competitive restaurants.

Cost Approaches to Pricing

- Another approach, which the authors believe is the best way to examine menu pricing, is to view it primarily from a *cost approach to pricing*.
- The best methods used by restaurateurs to set prices consider an operation's costs and profit goals when determining menu prices. Currently, the two most popular pricing systems are those that are based upon:
    - Food cost percentage
    - Item contribution margin

Food Cost Percentage

- The formula for computing food cost percentage for a restaurant is as follows:

$$\frac{\text{Cost of Food Sold}}{\text{Food Sales}} = \text{Food Cost \%}$$

- This formula can be worded somewhat differently for a single menu item without changing its accuracy. Consider that:

$$\frac{\text{Item Food Cost}}{\text{Selling Price}} = \text{Item Food Cost \%}$$

- The principles of algebra allow you to rearrange the formula as follows:

$$\frac{\text{Item Food Cost}}{\text{Item Food Cost \%}} = \text{Selling price}$$

- This method of pricing is based on the idea that food cost should be a predetermined percentage of selling price. When you use a predetermined food cost percentage to price menu items, you are stating the belief that food cost in relationship to selling price is of vital importance. See *Go Figure!* in the text for an example of this.

- A second formula for arriving at appropriate selling prices based on predetermined food cost % goals can be employed. This method uses a cost factor or multiplier that can be assigned to each desired food cost percentage (see *Go Figure!* in the text):

$$\frac{1.00}{\text{Desired Item Food Cost \%}} = \text{Pricing Factor}$$

- A factor table for desired item food cost percentages from 20% to 45% is shown in Figure 7.2.

Go Figure!

A factor, when multiplied times the item's cost, will result in a selling price that yields the desired item food cost percentage. The computation would be as follows:

$$\text{Pricing Factor} \times \text{Item Food Cost} = \text{Selling Price}$$

Item Contribution Margin

- Some managers prefer an approach to menu pricing that is focused on an **item**

contribution margin, defined as the amount that remains after the food cost of a menu item is subtracted from that item's selling price.
- Item contribution margin, then, is the money that "contributes" to paying for labor and other expenses *and* providing a profit.
- Some restaurateurs refer to item contribution margin as **item gross profit margin** (selling price minus item food cost). This term is sometimes used because it employs the same calculation as gross profit margin on the income statement (food sales minus food cost).

Go Figure!

The item contribution margin is computed as follows:

Selling Price – Item Food Cost = Item Contribution Margin

The selling price is calculated as follows:

Item Food Cost + Desired Item Contribution Margin = Selling Price

- Managers who rely on the contribution margin approach to pricing do so in the belief that the average contribution margin per item is a more important consideration in pricing decisions than food cost percentage.

Food Cost Percentage vs. Item Contribution Margin

- For the average managerial accountant, understanding the use of food cost percentage, item contribution margin, or a combination of both will enable him or her to arrive at appropriate pricing decisions.
- Pricing should be viewed as but an important process with an end goal of establishing a good price/value relationship in the mind of your guest while achieving profits for your operation.
- It is important that the menu not be priced so low that no profit is possible or so high that you will not be able to sell a sufficient number of items to make a profit.

Menu Price Analysis

- The best method of analyzing the profitability of a menu and its pricing structure should simply seek to answer the question, "How does the sale of this menu item contribute to the overall success of my operation?"
- Menu analysis involves marketing, imaging, sociology, psychology, and many times, the manager's emotions. Guests respond not just to weighty financial analyses, but rather to menu design, the description of the menu item, the placement of items on the menu, their price, and their current popularity.

- The menu analysis methods that have been widely used each seek to perform the analysis using one or more of the following important operational variables:
  - Food cost percentage
  - Popularity (sales mix)
  - Contribution margin
  - Selling price
  - Variable costs
- The most popular systems of menu analysis (shown in Figure 7.3) represent the three major philosophical approaches to menu analysis.
- The items and information related to menu items' cost, selling price, contribution margin, and popularity are compiled in a Menu Analysis Worksheet (refer to Figure 7.4). These figures can be used to illustrate the matrix analysis of food cost percentage and contribution margin, and goal value analysis. (See *Go Figure!* in the text to see how the rows of the Menu Analysis Worksheet are calculated.)

Matrix Analysis

- **Matrix analysis** is essentially an easy method used to make comparisons among menu items. A matrix allows menu items to be placed into categories based on their unique characteristics such as food cost %, popularity, and contribution margin.

Food Cost Percentage

- Matrix analysis that focuses on food cost percentage is the oldest and most traditional method used. Until the mid-1980s, it was overwhelmingly the single most popular method of evaluating the effectiveness of menu pricing decisions.
- When analyzing a menu using the food cost percentage method, you are seeking menu items that have the effect of *minimizing your overall food cost percentage*, since a lowered food cost percentage leaves more of the sales dollar to be spent for other operational expenses.
- A criticism of the food cost percentage approach is that items that have a higher food cost percentage may be removed from the menu in favor of items that have a lower food cost percentage but may also contribute fewer dollars to overall profit.
- To analyze a menu using the food cost percentage method, menu items must be segregated based on the following two variables:
  - Popularity (number sold)
  - Food cost percentage

Go Figure!

To determine average popularity (number sold) of menu items, the total number of items sold is divided by the number of items on the menu. In this case, the computation is:

$$\frac{\text{Total \# Sold}}{\text{Number of Menu Items}} = \text{Average \# Sold}$$

To determine the average item food cost percentage, the average total food cost is divided by average total sales. The computation is:

$$\frac{\text{Average Total Food Cost}}{\text{Average Total Sales}} = \text{Weighted Average Food Cost \%}$$

- With an average popularity of 100 covers sold per menu item per week, any item which sold more than 100 times would be considered *high* in popularity, while any item selling less than 100 times would be considered *low* in popularity.
- Similarly, with an overall average food cost of 35%, any menu item with a food cost percentage above 35% would be considered *high* in food cost percentage, while any menu item with a food cost below 35% would be considered *low*.
- The food cost percentage matrix to analyze these variables follows:

Popularity

		Low	High
Food Cost %	**High**	*Square 1* High Food Cost %, Low Popularity	*Square 2* High Food Cost %, High Popularity
	Low	*Square 3* Low Food Cost %, Low Popularity	*Square 4* Low Food Cost %, High Popularity

- See Figure 7.4 for sample data which can be used to create the actual matrix.
- In matrix analysis, each menu item inhabits one, and only one, square. When developing a menu that seeks to minimize food cost percentage, items in the fourth square are highly desirable. They should be well promoted and have high menu visibility.
- The characteristics of the menu items that fall into each of the four matrix squares are unique and, thus, should be managed differently. Each of the menu items that fall in the individual squares requires a special marketing strategy, depending on their square location. These strategies can be summarized as shown in Figure 7.5.
- The food cost percentage method is fast, logical, and time tested.
- If you achieve too high a food cost percentage, you run the risk that not enough money will remain to generate a profit on your sales. Again, however, you should be cautioned against promoting low-cost items with low selling prices at the expense of higher food percentage items with higher prices that may contribute greater gross profits. (For a discussion of this, see *Go Figure!* in the text.)

Contribution Margin

- When analyzing a menu using the contribution margin approach (also widely known as **menu engineering**), the operator seeks to produce a menu that *maximizes the menu's overall contribution margin.*
- Each menu item will have its own contribution margin, defined as the amount that remains after the food cost of the item is subtracted from the item's selling price.
- Contribution margin is the amount that you will have available to pay for your labor and other expenses and to keep for your profit.
- A common, and legitimate, criticism of the contribution margin approach to menu analysis is that it tends to favor high-priced menu items over low-priced ones, since higher priced menu items, in general, tend to have the highest contribution margins. Over the long term, this can result in sales techniques and menu placement decisions that tend to put in the guest's mind a higher check average than the operation may warrant or desire.
- To analyze a menu using the contribution margin method, menu items must be segregated based on the following two variables:
    - Popularity (number sold)
    - Contribution margin

Go Figure!

In Figure 7.4, to determine average popularity (number sold) of the menu items, the total number of items sold is divided by the number of items on the menu. In this case, the computation is:

$$\frac{\text{Total \# Sold}}{\text{Number of Menu Items}} = \text{Average \# Sold}$$

To determine the weighted average item contribution margin, the average total contribution margin is divided by the average # sold. The computation is:

$$\frac{\text{Average Total Contribution Margin}}{\text{Average \# Sold}} = \text{Weighted Average Item Contribution Margin}$$

- The contribution margin matrix is developed along much the same lines as the food cost percentage matrix.
- The contribution margin matrix to analyze these variables follows:

<table>
<tr><td rowspan="3">Contribution Margin</td><td colspan="4" align="center">Popularity</td></tr>
<tr><td></td><td>Low</td><td>High</td></tr>
<tr>
<td>High</td>
<td>*Square 1*
High Contribution Margin, Low Popularity</td>
<td>*Square 2*
High Contribution Margin, High Popularity</td>
</tr>
<tr>
<td colspan="1"></td>
<td>Low</td>
<td>*Square 3*
Low Contribution Margin, Low Popularity</td>
<td>*Square 4*
Low Contribution Margin, High Popularity</td>
</tr>
</table>

- See Figure 7.4 for sample data which can be used to create the actual matrix.
- Again, each menu item finds itself in one, and only one, matrix square. Using the contribution margin method of menu analysis, it is desirable to have as many of the menu items as possible to fall within square 2, reflecting high contribution margin and high popularity.
- A manager would seek to give high menu visibility to items with high contribution margin and high popularity when using the contribution margin approach.
- Each of the menu items that fall in the four squares requires a special marketing strategy, depending on its location. These strategies are summarized in Figure 7.6.
- The selection of either food cost percentage or contribution margin as a menu analysis technique is really an attempt by the foodservice operator to answer the following questions:
  o Are my menu items priced correctly?
  o Are the individual menu items selling well enough to warrant keeping them on the menu?
  o Is the overall profit margin on my menu items satisfactory?
- Some sophisticated observers feel that neither the matrix food cost nor the matrix contribution margin approach is tremendously effective in analyzing menus. Because the axes on the matrix are determined by the average food cost percentage, contribution margin, or sales level (popularity), some menu items will *always* fall into the less desirable categories. Eliminating the poorest items only shifts other items into undesirable categories. For an illustration of this drawback to matrix analysis, see *Go Figure!* in the text.
- One answer to address complex questions related to price, sales volume, and overall profit margin is to avoid the overly simplistic matrix analysis and employ a more effective method of menu analysis called Goal Value Analysis.

Goal Value Analysis

- In 1995, at the height of what was known as the **value pricing** (extremely low pricing strategies used to drive significant increases in guest counts) debate, goal value analysis proved its effectiveness.
- Essentially, **goal value analysis** is a menu pricing and analysis system that compares goals of the foodservice operation to performance of individual menu

- items. It uses the power of an algebraic formula to replace less sophisticated menu averaging techniques.
- The advantages of goal value analysis include ease of use, accuracy, and the ability to simultaneously consider more variables than is possible with two-dimensional matrix analysis.
- Both the food cost percentage and contribution margin approaches to menu analysis relate to the actual costs of food and beverages. Today, the cost of labor is likely to *exceed* that of food and beverages in many restaurants. In such a situation, how logical is it to utilize a menu analysis system that is based upon food costs or contribution margin alone, and that ignores labor costs?
- Goal value analysis evaluates each menu item's food cost percentage, contribution margin, popularity, and, *unlike* the two previous analysis methods introduced, includes the analysis of the menu item's nonfood variable and fixed costs as well as its selling price.
- The total dollar amount of **fixed costs** does not vary with sales volume, while the total dollar amount of **variable costs** changes as volume changes. An example of a fixed cost is manager salaries, and an example of a variable cost is hourly wages.
- Returning to the data in Figure 7.4, we see an overall food cost % of 35%. In addition, 700 guests were served at an entrée check average of $16.55. If we knew the overall fixed and variable costs, we would know more about the profitability of each of the menu items.
- One difficulty, of course, resides in the assignment of nonfood variable costs to individual menu items. *The majority of nonfood variable costs assigned to menu items is labor cost.*
- The *variable* labor cost of preparing many dishes is very different. For analysis purposes, most operators find it convenient to assign nonfood variable costs to individual menu items based on the overall restaurant's nonfood variable costs.
- Total variable costs are all the costs that vary with sales volume, excluding the cost of the food itself. Those variable costs are computed from the income statement. If they account for 30% of the total sales, a variable cost of 30% of selling price would be assigned to each menu item.
- Having compiled the information in Figure 7.4, the algebraic goal value formula can be used to create a specific goal value for the entire menu, and the same formula can also be used to compute the goal value of each individual menu item.
- Menu items that achieve goal values higher than that of the overall menu goal value will contribute greater than average profit percentages. As the goal value for an item increases, so, too, does its profitability.
- The overall menu goal value can be used as a "target" in this way, assuming that the average food cost %, average number of items sold per menu item, average selling price (check average), and average variable cost % all meet the overall profitability goals of the restaurant.
- The goal value formula is as follows:

$$A \times B \times C \times D = \text{Goal Value}$$

where
A = 1.00 − Food Cost %
B = Item Popularity
C = Selling Price
D = 1.00 − (Variable Cost % + Food Cost %)

- Note that *A* in the preceding formula is actually the contribution margin *percentage* of a menu item and that *D* is the amount available to fund fixed costs and provide for a profit after all variable costs are covered.

Go Figure!

Using 30% for variable cost % and the average/ weighted average numbers from Figure 7.4 for # sold (100), selling price ($16.55), and food cost % (35%), the formula can be calculated to compute the goal value of the *total menu* as follows:

A	× B	× C	× D	= Goal Value
(1.00 − 0.35)	× 100	× $16.55	× [1.00 − (0.30 + 0.35)]	= Goal Value
		or		
0.65	× 100	× $16.55	× 0.35	= 376.5

According to this formula, any menu item whose goal value equals or exceeds 376.5 will achieve profitability that equals or exceeds that of the overall menu.

- The computed goal value is neither a percentage nor a dollar figure because it is really a numerical target or score. (See Figure 7.7 for an example of the goal value data needed to complete a goal value analysis.)
- Refer to Figure 7.8 for the results of the goal value analysis. Should items that fall substantially below the overall goal value score be replaced? The answer, most likely, is no *if* the manager is satisfied with the current target food cost percentage, profit margin, check average, and guest count. Every menu will have items that are more (and less) profitable than others.
- Many operators develop and promote items called loss leaders. A **loss leader** is a menu item that is priced very low, sometimes even below total costs, for the purpose of drawing large numbers of guests to the operation, while their fellow diners may order items that are more profitable.

- The accuracy of goal value analysis is well documented. Used properly, it is a convenient way for management to make decisions regarding required profitability, sales volume, and pricing.
- Items that do not achieve the targeted goal value tend to be deficient in one or more of the key areas of food cost percentage, popularity, selling price, or variable cost percentage. In theory, all menu items have the potential of reaching the goal value. Refer to *Go Figure!* in the text to see an example of goal value analysis of a menu item. This item did not meet the goal value target. Why? There can be several answers.
- One is that the item's food cost % is too high. This can be addressed by reducing portion size or changing the item's recipe since both of these actions have the effect of reducing the food cost % and, thus, increasing the *A* value.
- A second approach to improving the goal value score is to work on improving the *B* value, that is, the number of times the item is sold. This may be done through merchandising or incentives to service staff for upselling this item.
- Variable *C*, menu price, can also be adjusted upward; however, adjustments upward in *C* may well result in declines in the number of items sold (*B* value)!
- Increases in the menu price will also have the effect of *decreasing* the food cost % and the variable cost % of the menu item (and increasing the contribution margin). An easy way to determine the effects of changes made to goal values is to use an Excel spreadsheet.
- Sophisticated users of the Goal Value Analysis system can modify the formula to increase its accuracy and usefulness even more.
- Goal value analysis will also allow you to make better decisions more quickly. Anytime you determine a desired goal value *and* when any three of the four variables contained in the formula are known, you can solve for the fourth unknown variable by using goal value as the numerator and placing the known variables in the denominator (see Figure 7.9).
- To illustrate how the data in Figure 7.9 can be used, see *Go Figure!* in the text.
- Goal value analysis is becoming increasingly linked to break-even analysis because of their mathematical similarities (see Chapter 9).
- In addition to the ability to analyze multiple cost variables simultaneously, goal value analysis is valuable because it is not, as is matrix analysis, dependent on *past* operational performance to establish profitability. It can be used by management to establish *future* menu targets. An illustration of this concept can be seen in *Go Figure!* in the text.
- Each item on next year's menu should be evaluated with the new goal value in mind. Actual profitability will be heavily influenced by sales mix, so all pricing, portion size, and menu placement decisions become critical.
- Remember, however, that a purely quantitative approach to menu analysis is neither practical nor desirable. Menu analysis and pricing decisions are always a matter of experience, skill, and educated predicting because it is difficult to know in advance how changing any one menu item may affect the sales mix of the remaining items.

Key Terms & Concepts Review

Match the key terms with their correct definitions.

1. Sales mix _____
2. Supply and demand _____
3. Price/value relationship _____
4. Item contribution margin _____
5. Item gross profit margin _____
6. Matrix analysis _____
7. Menu engineering _____
8. Value Pricing _____
9. Goal value analysis _____
10. Fixed costs _____
11. Variable costs _____
12. Loss Leader _____

a. An extremely low pricing strategy used to drive significant increases in guest counts.
b. A menu pricing and analysis system that compares goals of the foodservice operation to performance of individual menu items.
c. A method in which the operator seeks to produce a menu that maximizes the menu's overall contribution margin.
d. Costs that increase as sales volume increases and decrease as sales volume decreases.
e. A menu item that is priced very low, sometimes even below cost, for the purpose of drawing large numbers of guests to the operation.
f. The economic law that states that, for most products purchased by consumers, as the price of an item increases, the number of those items sold will generally decrease. Conversely, as the price of an item decreases, the number of those items sold will generally increase.
g. Costs that remain constant despite increases or decreases in sales volume.
h. A reflection of guests' view of how much value they are receiving for the price they are paying.
i. A method used to make comparisons among menu items which places them into categories based on their unique characteristics such as food cost %, popularity, and contribution margin.
j. The frequency with which specific menu items are selected by guests.
k. The amount that remains after the food cost of a menu item is subtracted from that item's selling price.
l. Also referred to as Item contribution margin.

Discussion Questions

1. List the factors that influence a foodservice operation's menu price.
   _____
   _____
   _____

2. List and explain the formulas for computing food cost percentage and item contribution margin.
   _____
   _____
   _____

3. Describe the matrix analysis method, using food cost percentage.
   _____
   _____
   _____

4. Explain goal value analysis and list each of the values used.
   _____
   _____
   _____

Quiz Yourself
Choose the letter of the best answer to the questions listed below.

Questions 1 – 10 use the following information taken from the operating and financial reports of Rosie's Chicken Palace, owned by Rosie Mendez:

Aunt Jane's Chicken Dinner price	$10.00
Food cost for Aunt Jane's Chicken Dinner plate	$3.50
# of Aunt Jane's Chicken Dinners sold weekly	150
Rosie's Total Cost of Food Sold	$4,800
Rosie's Total Food Sales for the week	$15,000
Total Meals Sold for the week	1700
Total Items on Menu	10
Rosie's Desired Food Cost %	30%

1. Calculate Rosie's total revenue this week from Aunt Jane's Chicken Dinner.
   a. $1,500
   b. $ 480
   c. $ 975
   d. $ 525

2. Rosie is considering raising the price on Aunt Jane's Chicken Dinner to $12. If she does, how many will she have to sell to maintain the current revenue for this item?
   a. 150
   b. 429
   c. 125
   d. 175

3. What is the food cost percentage for Rosie's Chicken Palace for the week?
   a. 35%
   b. 32%
   c. 10%
   d. 9%

4. What is the food cost percentage for Aunt Jane's Chicken Dinner?
   a. 35%
   b. 32%
   c. 10%
   d. 9%

5. Rosie is considering using a pricing factor to adjust the prices on her menu to reflect her desired food cost %. What pricing factor should she use?
   a. 2.857
   b. 3.133
   c. 9.999
   d. 3.333

6. If Rosie uses the pricing factor based on her desired food cost % to price Aunt Jane's Chicken Dinner, what would her new price be for this popular menu item?
   a. $10.00
   b. $10.96
   c. $11.67
   d. $12.50

7. What is the current item contribution margin for Aunt Jane's Chicken Dinner?
   a. $7.00
   b. $6.50
   c. $6.00
   d. $3.50

8. Calculate the average popularity (number sold) of the menu items at Rosie's.
   a. 170
   b. 150
   c. 17
   d. 15

9. Compute the Overall Menu Goal Value for Rosie's Chicken Palace, assuming a variable cost percentage of 30% and a desired average selling price of $9.00.
   a. 137.7
   b. 661.5
   c. 428.4
   d. 735.0

10. Compute the goal value for Aunt Jane's Chicken Dinner at the current price, assuming a variable cost percentage of 30%.
    a. 137.70
    b. 420.00
    c. 476.00
    d. 341.25

Chapter 8

Revenue Management for Hotels

Highlights

* Establishing Room Rates
  - The Hubbart Room Rate Formula
  - The $ 1.00 per $ 1,000 Rule
  - Alternative Room Rate Methodologies
  - Web-Influenced Room Rate Methodologies
* Revenue Management
  - Net ADR Yield
  - Flow-Through
  - GOPPAR
* Non-Room Revenue
* Key Terms and Concepts Review
* Discussion Questions
* Quiz Yourself

Study Notes

Establishing Room Rates

- Any serious exploration of hotel room rates and their management must include basic information about room rate economics. **Room rate economics** recognizes that, when the supply of hotel rooms is held constant, an increase in demand for those rooms will result in an increase in their selling price. Conversely, when supply is held constant, a decrease in demand leads to a decreased selling price.
- Understanding the law of demand is critical because, unlike managers in other industries, hoteliers cannot increase their inventory levels of rooms (supply) in response to increases in demand.
- Hotel managers must also understand that their own inventory of rooms is highly perishable. If a hotel does not sell room 101 on Monday night, it will never again be able to sell that room on that night, and the potential revenue that would be generated from the sale is lost forever.
- Since information about supply is readily known, and since forecast data helps to estimate demand, you can learn to accurately gauge the relationship between guestroom supply and demand. Using this information, you can determine the best rates to be assigned to each of your rooms.
- A **rack rate** is the price at which a hotel sells its rooms when no discounts of any kind are offered to the guests. Rack rates, however, will vary based upon the type

of room sold. Figure 8.1 lists the rack rates that are associated with Paige Vincent's Blue Lagoon Water Park Resort based on her **room mix** (the variety of room types) in her hotel.

- In Figure 8.1, rack rates vary by bed type, by amenities, by location and by size.
- Some hotels have very strong seasonal demand. These hotels will have a **seasonal rate** that is higher or lower than the standard rack rate and that is offered during that hotel's highest volume season.
- In some cases, it makes sense for hoteliers to create **special event rates**. Sometimes referred to as "super" or "premium" rack, these rates are used when a hotel is assured of very high demand levels (e.g., Mardi Gras in New Orleans and New Year's Eve in New York City).
- Hotels often negotiate special rates for selected guests. In most cases, these negotiated rates will vary by room type. In addition to rack and negotiated rates, hotels typically offer **corporate rates, government rates, and group rates.**
- Some hotels have great success "packaging" the guest rooms they sell with other hotel services or local area attractions. When a hotel creates a package, the **package rate** charged must be sufficient to ensure that all costs associated with the package have been considered.
- A hotel's revenue managers can also create discounts at various percentage or dollar levels for each rate type we have examined. The result is that a hotel, with multiple room types and multiple rate plans, may have literally *hundreds* of rates types programmed into its property management system.
- A **property management system (PMS)** is a computer system used to manage guest bookings, online reservations, check-in/check-out, and guest purchases of amenities offered by the hotel.
- In addition, the use of one or more authorized **fade rates**, a reduced rate authorized for use when a guest seeking a reservation is hesitant to make the reservation because the price is perceived as too high, can result in even more room rates to be managed.

The Hubbart Room Rate Formula

- Hoteliers want to maximize their profits and thus collect the highest rate possible for their rooms. However, the rate cannot be so high that it discourages guests from staying at the hotel, nor can it be so low that it prevents the hotel from making a profit.
- The room rate charged should not result from a mere "guess" about its appropriateness but, ideally, should evolve from a rational examination of guest demand (because it is the most significant factor impacting room rates) *and* a hotel's costs of operation with specific and accurate assumptions.
- Recognized by hoteliers world-wide, the Hubbart Room Rate Formula for determining room rates was developed in the mid-1950s by the national hotel accounting firms of Horwath & Horwath and Harris Kerr Forster.
- The **Hubbart formula** is used to determine what a hotel's average daily rate (ADR) *should* be to reach the hotel owner's financial goals.

- To compute the Hubbart formula, specific financial and operational assumptions are determined. These include dollar amounts for property construction (or purchase), the total cost of the hotel's operations, the number of rooms to be sold, and the owner's desired ROI on the hotel's land, property, and equipment.
- The Hubbart formula is a "bottom-up" approach because it literally requires you to completely reverse the income statement from the bottom up (see Figure 8.2 for the comparison between the normal format of the income statement and the bottom-up format for the Hubbart formula).
- Using the bottom-up approach, you start by calculating the desired net income based on the owner's desired return on investment (ROI) and work your way up the income statement by adding back estimated taxes, non-operating expenses, and undistributed operating expenses and then subtracting out estimated operated departments income (excluding rooms). The result will be the estimated operated department income for rooms.
- Next, you can separate the estimated operated department income for rooms into rooms revenue and rooms expenses. Once rooms expenses are subtracted out, rooms revenue will remain. This revenue can then be split again to determine number of rooms to be sold and, finally, ADR.
- The resulting ADR is the average price that should be charged for your rooms in order to achieve the owner's desired net income (ROI).
- To illustrate the Hubbart formula, the Blue Lagoon Water Park Resort's Income Statement is shown in Figure 8.3. Assume for the sake of calculating this formula that we *do not* know the operated department income for rooms, rooms revenues, or rooms expenses. Remember, the point of using the Hubbart formula is to *predict* rooms revenue, and subsequently, ADR.
- For a detailed analysis of the Hubbart formula, see *Go Figure!* in the text.
- For a summary of the Hubbart formula calculations for the Blue Lagoon Water Park Resort, see Figure 8.4.
- The seven steps required to compute the Hubbart formula are summarized in Figure 8.5.
- The Hubbart formula is useful because it requires managerial accountants and hoteliers to consider the hotel owner's realistic investment goals and the costs of operating the hotel before determining the room rate.
- The formula has been criticized for relying on assumptions about the reasonableness of an owner's desired ROI and the need to know expenses that are affected by the quality of the hotel's management. Another criticism is that the formula requires the room rate to compensate for operating losses incurred by other areas (such as from telecommunications).
- The formula's primary shortcoming may relate to identifying the number of rooms forecasted to be sold. The number of rooms sold is dependent, to a significant degree, on the rate charged for the rooms. However, the Hubbart formula requires that the number of rooms sold be estimated *prior* to knowing the rate at which they would sell.
- Despite its limitations, the Hubbart formula remains an important way to view the necessity of developing a room rate that:
  o Provides an adequate return to the hotel's owner(s)

- o Recovers the hotel's non-operating expenses
- o Considers the hotel's undistributed operating expenses
- o Accounts for all the hotel's non-room operated departments income (or loss)
- o Results in a definite and justifiable overall ADR goal

The $1.00 per $1,000 Rule

- One alternative way that hoteliers have historically determined room rate is the **$1.00 per $1,000 rule.** This rule states that, for every $1,000 invested in a hotel, the property should charge $1.00 in ADR.
- Advocates defend the $1.00 per $1,000 rule of thumb because areas in which building or purchase costs are higher tend to be the areas where ADRs can also be higher.
- The dollar-per-thousand rule is most accurate for hotels that have high occupancies, high ADRs for their area of operation, and are newly built. On the other hand, large, old properties frequently fail to achieve the dollar-per-thousand standard.
- Despite some limitations, the $1.00 per $1,000 rule does reflect the tendency for hotel buyers to discuss hotel selling prices in terms of a hotel's **cost per key,** which is the average purchase price of a hotel's guestroom expressed in thousands of dollars. Cost per key is also frequently called **average cost per room**.

Go Figure!

Average Cost per Room is calculated as follows:

$$\frac{\text{Purchase Price}}{\text{Number of Rooms}} = \text{Average Cost per Room}$$

Then, ADR is calculated as follows:

$$\frac{\text{Average Cost per Room}}{\$1,000} = \text{ADR}$$

- It is important to recognize that the rate computed using the $1.00 per $1,000 rule does *not* become the hotel's rack rate. Instead, it is the overall ADR that the hotel must achieve when its sells all of its various rooms at all of their respective rates.

Alternative Room Rate Methodologies

- Additional historical methods of rate determination include those based upon the square footage of guestrooms and rates determined by various "ideal" sales levels of the different hotel room types available to be sold.

- However, today's hotel room rate structures have been changed, and changed forever, by the advent of the Internet as the most popular method used for selling hotel rooms.

Web-Influenced Room Rate Methodologies

- Properly pricing hotel rooms is critical to attracting first time and repeat business. However, close examination of many pricing tactics would reveal that they often use one or more of the following non-traditional, non-cost methods to establish rates:
    - **Competitive Pricing.** Charge what the competition charges.
    - **Follow the Leader Pricing.** Charge what the dominant hotel in the area charges.
    - **Prestige Pricing.** Charge the highest rate in the area and justify it with better product and/or service levels.
    - **Discount Pricing.** Reduce rates below that of the likely competitors.
- As a result of the Internet, consumers can easily compare prices, but so can a hotel's major competitors. Gone are the times when night auditors or others on the front office staff conducted the nightly **call-around** to ask other hotels' night auditors what their hotels were charging for rooms and then used that information (often of questionable accuracy) to make decisions about what their own hotel's rate offerings should be.
- While the call-around was standard practice as late as the early 2000s, consider modern hoteliers utilizing one of the many websites similar to http://www.Travelaxe.com and others that allow him/her to easily:
    - Select competitive hotels whose rates are to be monitored
    - Obtain real-time room rates offered by these hotels on any number of travel websites advertising the rates
    - Search the rates and sites as frequently as desired
    - Perform rate comparisons by specific check-in and check-out dates
    - Assess rate comparisons based upon room type
    - Assess rate comparisons based upon date of guest arrival
- Guests care very little how much it "costs" a hotel to provide its rooms. They care about the lodging value they receive. As a result, a hotel's rates are heavily influenced by the laws of supply and demand. If, on a given Saturday, all similar hotels in a market area offer guest rooms in the range of $100 - $150 per night, it would be difficult for a single hotel of the same type to command a rate of $250 per night even if its operating costs justify this rate.
- The rate at which a hotel first sells its rooms to guests may not be the rate those guests will ultimately pay. A guest can make a hotel reservation at a given rate, and, every day until the date of arrival, can go online to shop for an even lower price for the same room. If a lower rate were to be found, the guest could re-contact the hotel, cancel the original reservation, and secure the new, lower rate.

Revenue Management

- **Revenue management**, also called **yield management**, is a set of techniques and procedures that use hotel specific data to manipulate occupancy, ADR, or both for the purpose of maximizing the revenue yield achieved by a hotel.
- **Yield** is a term used to describe the percentage of total potential revenue that is actually realized. **Revenue managers** are responsible for making decisions regarding the pricing and selling of guest rooms in order to maximize yield.

Go Figure!

A hotel's yield would be calculated as follows:

$$\frac{\text{Total Realized Revenue}}{\text{Total Potential Revenue}} = \text{Yield}$$

- RevPAR is a combination of ADR and occupancy % and is calculated using the following formula:

$$\text{ADR} \times \text{Occupancy \%} = \text{RevPAR}$$

- To increase yield simply means to increase the hotel's RevPAR. Therefore, any change (decrease or increase) in either or both of the factors comprising RevPAR will change the yield of the hotel's revenue.
- Because hotel rooms are highly perishable, the goal of revenue management is to consistently maintain the highest possible revenue from a given amount of inventory.
- Revenue management techniques are used during periods of low, as well as high, demand.
- Although the actual revenue management techniques used by hoteliers vary by property, in their simplest form, all these techniques are employed to:
  - Forecast demand
  - Eliminate discounts in high demand periods
  - Increase discounts during low demand periods
  - Implement "Special Event" rates during periods of extremely heavy demand
- Sophisticated mathematical programs that help hoteliers manage revenues are built into most property management systems (PMS) used in the industry today. Using information gleaned from the hotel's historical sales data, revenue management features in a PMS can:
  - Recommend room rates that will optimize the number of rooms sold
  - Recommend room rates that will optimize sales revenue
  - Recommend special room restrictions that serve to optimize the total revenue generated by the hotel during a specific time period

- o Identify special high consumer demand dates that deserve special management attention to pricing
- PMS systems can "remember" more important dates than can an individual hotelier. However, it is hotelier's skill and experience that is most critical to the revenue maximization process. The goal of a talented revenue manager is to increase RevPAR, not only on a daily basis, but on a long-term basis as well.
- In the hotel industry, a **competitive set (comp set)** consists of those hotels with whom a specific hotel competes and to which it compares its own operating performance.
- To fully evaluate RevPAR changes, hoteliers look to the relative performance of their comp set. They do so to better understand the room rate economics that affected their own property during a specific time period, as well as how the hotel's management responded to the supply and demand challenges they faced during that period.
- To better understand the shortcomings of an over-emphasis on RevPAR, it is important to take a closer look at its two fundamental components, occupancy % and ADR.

Occupancy %

- It might seem that the occupancy percentage for a hotel would be a straightforward calculation. A room revenue statistics report generated from the Blue Lagoon Water Park Resort's property management system produced the information needed to calculate their occupancy percentage (refer to Figure 8.6).

Go Figure!

Exactly how should Paige at the Blue Lagoon Water Park Resort compute her occupancy percentage? Should she:

1. Include only sold rooms in her computation? If so, her formula would be:

$$\frac{\text{Rooms Sold}}{\text{Total Rooms in Hotel}} = \text{Occupancy \%}$$

2. Include complimentary rooms as well as sold rooms in her computation? If so, her formula would be:

$$\frac{\text{Rooms Sold + Comp Rooms Occupied}}{\text{Total Rooms in Hotel}} = \text{Occupancy \%}$$

3. Subtract non-sellable out-of-order rooms from her rooms available count? If so, her formula would be:

$$\frac{\text{Rooms Sold}}{\text{Total Rooms in Hotel} - \text{OOO Rooms}} = \text{Occupancy \%}$$

4. Subtract non-sellable on change rooms from her rooms available count? If so, her formula would be:

$$\frac{\text{Rooms Sold}}{\text{Total Rooms in Hotel} - \text{On-Change Rooms}} = \text{Occupancy \%}$$

ADR

- Average daily rate (ADR) is also a critical component of RevPAR. Generally, hotel managers calculate ADR using one of the two following formulas:

$$\frac{\text{Total Rooms Revenue}}{\text{Total Number of Rooms Sold}} = \text{ADR}$$

or

$$\frac{\text{Total Rooms Revenue}}{\text{Total Number of Rooms Occupied}} = \text{ADR}$$

- Notice that the only difference in the two formulas are the words *sold* and *occupied*. See *Go Figure!* in the text for further illustration of the difference between these two formulas for calculating ADR.
- Despite the slight differences in these two ADR computations, neither is as useful to the hotel's owners and managers as the computation of the net ADR yield.

Net ADR Yield

- **Net ADR Yield** is the percentage of ADR actually received by a hotel *after* subtracting the cost of fees and assessments associated with the room's sale. For a single room it is computed as:

$$\frac{\text{Room Rate - Reservation Generation Fees}}{\text{Room Rate Paid}} = \text{Net ADR Yield}$$

- To really understand net ADR yield, you must first understand how hotel rooms were sold in the past as well as how they are sold in today's competitive marketing environment. In the distant past, hotels clearly preferred that guests arrive with a previously made reservation. Of course, if the hotel had vacant rooms, the front office agent would quote a rate (often higher than that quoted to other guests) to the non-reserved guest and the room would be sold.

- Today, most hotel guests already have a room reservation prior to arrival, and the **reservation distribution channels** (sources of reservations) used to make their reservations will charge the hotel widely varying fees for making them.
- When a guest makes a reservation via the Internet, no less than three reservation-generation fees are typically charged to the hotel, including fees from:
    - **The Internet Travel Site** - website for booking travel to end-users
    - **The Global Distribution System (GDS)** - system that books and sells rooms for multiple companies
    - **The Central Reservation System (CRS)** – system used by companies to centrally book reservations
- Consider the information in Figure 8.7. As you can see, the hotel pays "zero" reservation fees on a walk in reservation and pays several fees for the reservation made by the Internet user.
- If the reservation is made by a travel agent, additional fees are charged.
- With increased usage of high priced distribution channels, a room's selling price (quoted) and the ADR the hotel actually receives can be radically different.
- For a further illustration of net ADR yield, see *Go Figure!* in the text.
- Clearly, it is the ADR *after* the cost of reservation generation fees that should be most important to hoteliers and their attempts to increase RevPAR. If net ADR yield is not used, hotel owners and managers run the risk of significantly over inflated RevPARs accompanied by significantly reduced profits as well.

Flow-Through

- **Flow-through** is a measure of the ability of a hotel to convert increased revenue dollars to increased gross operating profit dollars. Consider the simplified income statements detailing revenue and expenses for January 2010 and for the same month of the prior year for the Blue Lagoon Water Park Resort (see Figure 8.8).

Go Figure!

From Figure 8.8, the Blue Lagoon's flow-through for January 2010 is calculated as the change in gross operating profit (GOP) from the prior year divided by the change in total revenues from the prior year as follows:

$$\frac{\text{GOP This Year} - \text{GOP Last Year}}{\text{Total Revenues This Year} - \text{Total Revenues Last Year}} = \text{Flow-Through}$$

- Gross operating profit (GOP) is, in effect, total hotel revenue less those expenses that are considered directly controllable by management. Flow-through was created by managerial accountants to measure the ability of a hotel to covert increases in revenue directly to increases in GOP.
- When flow-through is high (over 50%), it reflects efficiency on the part of management in converting additional revenues into additional profits. For most

hotels, flow-throughs that are less than 50% indicate inefficiency in converting additional revenues into additional profits.

GOPPAR

- **Gross operating profit per available room (GOPPAR)** is defined as a hotel's total revenue minus its management's controllable expenses per available room. For example, the costs of a hotel's lawn care services, utility bills, and even food and beverage expenses are considered when computing GOPPAR. These same expenses are not, of course, considered when computing RevPAR.
- In most cases, those managers directly responsible for revenue generation do not control the majority of costs used to compute GOPPAR. How did it become popular to suggest GOPPAR as a method of evaluating the decision making of those revenue managers?
- For a simple example of why this is so, consider a hotel that elects to launch a major advertising campaign in its local market, which costs $100,000 per month, and increases RevPAR by $15,000 per month. Despite the fact that RevPAR certainly did increase, the amount of money spent by the hotel to increase RevPAR exceeded, by far, the actual amount of the revenue increase. Clearly, the short-term effect on hotel profitability will be a negative one.
- Thus, there are still some pitfalls to be aware of when analyzing a hotel's performance based solely on RevPAR. For example, in those cases where room revenue accounts for only 50 to 60% of total revenue (as is the case in large convention hotels), RevPAR represents only half of the hotel's revenues and neglects to consider all other sources of incremental revenues.
- It is for shortcomings such as these that hoteliers now consider an analysis of a hotel's GOPPAR to be of such importance.

Go Figure!

GOPPAR is calculated as follows:

$$\frac{\text{Gross Operating Profit}}{\text{Total Rooms Available to Be Sold}} = \text{GOPPAR}$$

- GOPPAR, because it reflects the gross operating profits (not revenue) of a hotel, actually provides a clearer indication of overall performance than does RevPAR. RevPAR indicates the performance of a hotel in terms of rooms inventory sales and marketing, however, it provides no indication of how much money the hotel actually is, or should be, making.
- GOPPAR takes into consideration the cost containment and management control of the hotel and must be considered in any effective rooms pricing strategy. The difficulty is *not* that RevPAR is a poor measurement, but rather it is the fact that

RevPAR should not be the *only* measurement of a hotel's revenue manager's effectiveness.

Non-Room Revenue

- **Non-room revenue** is important to the managers of both limited service and full-service hotels. It is common for limited service hotels to generate 5-20% of their total revenue from non-room sources. In full-service hotels, the non-rooms revenue generated may range from 20-50% of total revenue. Non-room revenue on a hotel's income statement is attributed to one of the following categories:
  o Food
  o Beverage
  o Telecommunications
  o Garage and Parking
  o Golf Course
  o Golf Pro Shop
  o Guest Laundry
  o Health Center
  o Swimming Pool
  o Tennis
  o Tennis Pro Shop
  o Other Operated Departments
  o Rentals and Other Income
- Not every hotel will create revenues in every non-room revenue area.

Food and Beverage Revenue

- Food and beverage revenues typically make up the largest portion of a hotel's non-room revenue. The pricing of a hotel's food and beverage products is not identical to that of a restaurant, and in many cases, it is not even similar. A restaurant manager wants to financially support the *restaurant* itself through sales. In a properly managed hotel food and beverage (F&B) department, the department head wants to financially support the *hotel* through sales.
- The philosophical and practical differences between the two approaches are immense. For example, in many hotels, complementary breakfast is served to all overnight guests. The food and beverage department in such a property may be reimbursed for the "cost" of providing the breakfast, but the actual "sales" value of the breakfast, including a profit, would not likely be transferred.
- The amount of profit generated by a traditional restaurant is relatively easy to calculate. All revenue is generated from the sale of food and/or beverage products in the restaurant and all expenses normally will be clearly identified in the accounting records of the establishment.
- The process of assigning revenues and expenses applicable to the F&B department in a hotel is more difficult. For example, a holiday weekend package plan that includes one night's stay, dinner, and breakfast is sold to guests for one price. How should the revenue generated from the guests be split between room revenue and the F&B department?

- Consider applicable expenses also. How much, if any, of the salary paid to the hotel's general manager, controller, and other staff specialists along with other expenses including utilities, landscaping, and marketing, should be allocated between departments (including F&B) within the hotel? It is, then, difficult to compare the profitability of a restaurant directly with that of its F&B counterpart in a lodging property.
- Hotel F&B face significant challenges to profitability. For example, traditional restaurants are open at the times when the majority of their guests want to be served, while a hotel restaurant will most likely remain open for three meal periods daily to serve hotel guests. As a result, payroll related costs in hotel F&B departments tend to be higher than their restaurant counterparts.
- Food service is often viewed as an amenity to attract guests and to provide food and beverage alternatives to increase the hotel's revenues. The role of the F&B department is, appropriately, secondary to that of those departments that sell and service guest rooms. Experienced managerial accountants understand this and resist the temptation to aggressively and expensively seek to market the F&B operation to non-hotel guests living in the local area.
- In addition to the restaurants and lounges found in hotels, for those hotels with liquor licenses (many limited service hotels do not have them), income statements will be prepared that include revenue and expense detail one or more of the following categories:
  - Room Service
  - Banquets
  - Breakfast
  - Lunch
  - Dinner
  - Meeting room rental
  - Meeting room set-up and décor
  - Audio and Visual (A/V) equipment rental
  - Service Charges
- **Service charges** are properly reported as F&B income because, unlike a tip, a service charge is a mandatory addition to a guest's food and beverage bill. Typical service charge rates in full service hotels range from 15-25 % of the guest's total pre-tax food and beverage charges.
- The portion of the hotel's mandatory service charge that is not returned to employees is often considered a direct contribution toward F&B profits.

Telecommunications Revenue

- In the not so distant past, in-room telephone toll charges contributed a significant amount of money to a hotel's annual revenue. Today, however, the advent of cell telephones, and the reputation for excessive charges that has plagued hotels have lead to significant declines in this revenue source. Many hotel brands have had to reduce or even eliminate their local telephone charges.
- As a result, for most hoteliers, focus on the telephone department has shifted from a "pricing" concern to a "cost" accounting and management concern.

- When guests make telephone calls outside the hotel, it is in the best interest of the hotel to route those calls in a way that minimizes the hotel's cost.
- When a call is made, the hotel will, depending on the distance and length of the call, add a charge to the guest's folio to offset the cost of that call. The procedure for doing so involves programming the hotel's **call accounting system**, to generate telephone toll charges based upon:
    - Time of day
    - Call length
    - Call distance (local or long distance)
    - Use or non-use of international service providers (carriers)
- Note that these cost factors are the same factors that will ultimately affect the hotel's own monthly telephone bill. As a result, all of these should indeed be considered when establishing appropriate in-room guest telephone charges.
- An effective call accounting system, when **interfaced** (electronically connected) with the hotel's property management system will post these charges directly to the guest's folio and provide the documentation (call date, time and number that was called) required to justify the collection of these charges when the guest checks out.

Other Operated Departments Revenue

- Most hoteliers will encounter and need to fully understand the following "other" revenue generating areas or departments:
    - Pay-per-view movies
    - Pay-per-play in-room games
    - In-room safes
    - Internet access charges
    - Miscellaneous other income

Key Terms & Concepts Review

Match the key terms with their correct definitions.

1. Room rate economics _____
   a. A pricing method in which night auditors make decisions about their own room rates by asking other hotels' night auditors about their room rates.

2. Rack rate _____
   b. A bottom-up pricing formula used to determine what a hotel's average daily rate should be to reach the hotel owners' financial goals.

3. Room mix _____
   c. System used by companies to centrally book reservations.

4. Seasonal rate _____
   d. The percentage of total potential revenue that is actually realized.

5. Special event rate _____ e. A pricing method used to charge the highest rate in the area and justify it with better product and/or service levels.

6. Corporate rate _____ f. A measure of the ability of a hotel to convert increased revenue dollars to increased gross operating profit dollars.

7. Government rate _____ g. A pricing method used to charge what the competition charges.

8. Group rate _____ h. An abbreviation for Wireless Fidelity access.

9. Package rate _____ i. Rooms that are vacant but not yet cleaned.

10. Property management system (PMS) _____ j. The percentage of average daily rate actually received by a hotel after subtracting the cost of fees and assessments associated with the room's sale.

11. Fade rate _____ k. Revenue generated by a hotel that is not specifically room sales.

12. Hubbart formula _____ l. System that books and sells rooms for multiple companies.

13. $1.00 per $1,000 rule _____ m. A computer system used to manage guest bookings, online reservations, check-in/check-out, and guest purchases of amenities offered by a hotel.

14. Cost per key _____ n. The variety of room types in a hotel.

15. Average cost per room _____ o. Individual responsible for making decisions regarding the pricing and selling of guest rooms in order to maximize yield.

16. Competitive pricing _____ p. The negotiated special rate offered by hotels for corporate guests.

17. Follow the leader pricing _____ q. A hotel's total revenue minus its management's controllable expenses per available room.

18. Prestige pricing _____ r. Sources of reservations.

19. Discount pricing _____ s. The price at which a hotel sells its rooms when no discounts of any kind are offered to the guests.

20. Call around _____ t. Electronically connected.

21. Revenue management _____ u. Fees assessed to guests when they have a guaranteed reservation and neither cancel the reservation nor show up at the hotel on their expected date of arrival.

22. Yield management _____
23. Yield _____
24. Revenue manager _____
25. Competitive set (comp set) _____
26. On-change rooms _____
27. Net ADR yield _____
28. Reservation distribution channels _____
29. Internet Travel Site _____
30. Global Distribution System (GDS) _____
31. Central Reservation System (CRS) _____
32. Flow-through _____
33. Gross Operating Profit Per Available Room (GOPPAR) _____
34. Non-room revenue _____
35. Service charge _____

v. A service that allows for "free" long distance calls via computer.

w. Economic tenet that states when the supply of hotel rooms is held constant, an increase in demand for those rooms will result in an increase in their selling price. Conversely, a decrease in demand leads to a decreased selling price.

x. The negotiated special rate offered by hotels for guests in a group.

y. A pricing method used to reduce rates below that of likely competitors.

z. A mandatory addition to a guest's food and beverage bill.

aa. The negotiated special rate offered by hotels for government guests.

bb. A pricing method used to charge what the dominant hotel in the area charges.

cc. A reduced room rate authorized for use when a guest seeking a reservation is hesitant to make the reservation because the price is perceived as too high.

dd. A system that generates hotel telephone toll charges based upon time of day, call length, call distance, and use or non-use of international service providers.

ee. A room rate charged by a hotel that combines the room rate with the prices of other hotel services or local area attractions.

ff. Group of hotels with whom a specific hotel competes and to which it compares its own operating performance.

gg. A room rate that is higher or lower than the standard rack rate and that is offered during a hotel's highest volume season.

hh. Also referred to as Cost per key.

ii. A set of techniques and procedures that use hotel specific data to manipulate occupancy, average daily rate, or both for the purpose of maximizing the revenue yield achieved by a hotel.

36. Voice Over Internet Protocol (VOIP) _____
37. Call accounting system _____
38. Interfaced _____
39. Wi-Fi _____
40. Hot spot _____
41. No-show charges _____

jj. The average purchase price of a hotel's guestroom calculated by dividing the purchase price of a hotel by the number of hotel rooms.

kk. A pricing formula that uses a rule of thumb that for every $1,000 invested in a hotel, the hotel should charge $1.00 in average daily rate.

ll. A Wi-Fi area that allows for high-speed wireless Internet access.

mm. The room rate used when a hotel is assured of very high demand levels due to special holidays or events.

nn. Website for booking travel to end-users.

oo. Also referred to as Revenue management.

Discussion Questions

1. List the seven steps required to complete the Hubbart room rate formula.

2. Explain the $1 per $1,000 rule.

3. Define the concept of revenue management for a hotel.

4. Define GOPPAR and explain why it is a clearer indication of overall performance of a hotel than RevPAR.

Quiz Yourself

Choose the letter of the best answer to the questions listed below.

Questions 1 – 10 use the following information taken from the operating and financial reports of The Sleepy-Time Inn, a limited service hotel located near a major interstate highway intersection.

Original Purchase Price	$9,000,000
Total # Rooms	150
Gross Operating Profit (GOP) This Year	$2,275,300
Gross Operating Profit (GOP) Last Year	$2,105,100
Total Rooms Sold This Year	40,230
Total Rooms Revenue This Year	$3,814,250
Total Rooms Revenue Last Year	$3,550,180

1. Calculate the average cost per room (cost per key) at The Sleepy-Time Inn.
   a. $60,000
   b. $ 6,000
   c. $22,371
   d. $23,596

2. Using the $1 per $1,000 rule, what should be the property's target ADR?
   a. $ 40.00
   b. $ 60.00
   c. $223.71
   d. $235.96

3. If the total potential revenue for a fully booked Sleepy-Time Inn this year is $5,201,250, calculate the hotel's yield.
   a. 73.3%
   b. 57.8%
   c. 42.4%
   d. 12.9%

4. On March 1st, The Sleepy-Time Inn sold 105 rooms. What was the hotel's occupancy percentage, assuming all rooms were available for sale?
   a. 60%
   b. 70%
   c. 80%
   d. 90%

5. On March 1st, one wing of the hotel was being remodeled, and 24 rooms were out-of-order due to construction. Taking these rooms into account, what was the hotel's occupancy percentage?
   a. 53.3%
   b. 63.3%
   c. 73.3%
   d. 83.3%

6. Using the formula for rooms sold, calculate this year's ADR for The Sleepy-Time Inn.
   a. $129.28
   b. $224.00
   c. $ 94.81
   d. $ 70.00

7. Using the ADR from question 6 and the occupancy percentage from question 5, calculate RevPAR for The Sleepy-Time Inn.
   a. $ 78.98
   b. $106.60
   c. $107.69
   d. $186.59

8. Assume that a guest paid $100 for a room night at The Sleepy-Time Inn and total reservation fees of $20 were charged to the hotel. Calculate the net ADR yield for the room night.
   a. 90%
   b. 80%
   c. 40%
   d. 20%

9. Calculate the flow-through for The Sleepy-Time Inn.
   a. (64.5%)
   b. 59.3%
   c. 59.7%
   d. 64.5%

10. Calculate the annual GOPPAR for this year at The Sleepy-Time Inn, assuming all rooms were available for sale and 365 days in a year.
    a. $56.56
    b. $38.45
    c. $41.56
    d. $52.33

Chapter 9

Managerial Accounting for Costs

Highlights

* The Concept of Cost
* Types of Costs
* Cost/Volume/Profit (CVP) Analysis
* Key Terms and Concepts Review
* Discussion Questions
* Quiz Yourself

Study Notes

The Concept of Cost

- In the two previous chapters you learned about the importance of knowing your costs before establishing menu prices as well as before determining the selling prices of hotel guest rooms. The word cost is a popular one in business. In the hospitality industry you can, among other things:
    - Control costs
    - Determine costs
    - Cover costs
    - Cut costs
    - Measure rising costs
    - Eliminate cost
    - Estimate costs
    - Budget for costs
    - Forecast costs
- Given all the possible approaches to examining costs, perhaps the easiest way to understand them is to consider their impact on a businesses' profit. Recall that the basic profit formula was presented in Chapter 1 as follows:

> **Revenue – Expenses = Profit**

- Throughout this chapter, the term cost will be used interchangeably with the term expense. Expressed in another way, and substituting the word "costs" for "expenses", the formula becomes:

> **Revenue = Profit + Costs**

- As you can see, at any specific level of revenue, the lower a business's costs, the greater are its profits. This is true in both the restaurant and hotel industries and is why understanding and controlling costs are so important to successful hospitality managers.
- When you consider how businesses operate, it is easy to see why all costs cannot be viewed in the same manner. In fact, there are a variety of useful ways in which hospitality managers and managerial accountants can view costs and thus can better understand and operate their own businesses.

Types of Costs

- Not all costs are the same. As a result, cost accountants have identified several useful ways to classify business costs. Among the most important of these are:
  - Fixed and variable costs
  - Mixed costs
  - Step costs
  - Direct and indirect (overhead) costs
  - Controllable and non-controllable costs
  - Other costs:
    - Joint costs
    - Incremental costs
    - Standard costs
    - Sunk costs
    - Opportunity costs
- Not all business costs can be objectively measured. In fact, in some cases, cost can be a somewhat subjective matter. In many cases, identifying a cost in the hospitality business can be as simple as reviewing an invoice for purchased meats or produce. The cost of "talking with customers regarding invoice questions" is an example of an activity performed inside many hospitality companies and one for which a clear-cut cost cannot be so easily assigned.
- Cost accountants facing such issues can assign each hospitality employee's time to different activities performed inside a company. An accountant can then determine the total cost spent on each activity by summing up the percentage of each worker's time and pay that is spent on that activity. This process is called **activity based costing** and it seeks to assign objective costs to somewhat subjective items such as the payment for various types of labor as well as the even more subjective management tasks involved with planning, organizing, directing and controlling a hospitality business.
- Using activity based costing to examine expenses and thus better manage a business is called **activity based management** and it is just one example of how fully understanding costs can help you make better decisions and operate a more successful business.

Fixed and Variable Costs

- As a manager, some of the costs you incur will stay the same each month. For that reason they are called fixed costs. A **fixed cost** is one that remains constant despite increases or decreases in sales volume (number of guest or number of rooms). Typical examples of fixed costs include payments for insurance policies, property taxes, and management salaries.
- The relationship between sales volume and fixed costs is shown in Figure 9.1, where the cost in dollars is displayed on the y (vertical) axis, and sales volume is shown on the x (horizontal) axis. Note that the cost is the same regardless of sales volume.
- In some cases, the amount hospitality managers must pay for an expense will not be fixed but will vary based on the success of their business. A **variable cost** is one that increases as sales volume increases and decreases as sales volume decreases.
- The relationship between sales volume and variable cost is shown in Figure 9.2, where the cost in dollars is displayed on the y (vertical) axis, and sales volume is shown on the x (horizontal) axis.

Go Figure!

The total variable cost is computed as follows:

Variable Cost per Guest (VC/Guest) x Number of Guests = Total Variable Cost

If the total variable cost and the number of guests are known, VC/Guest can be determined. Using basic algebra, a variation of the total variable cost formula can be computed as follows:

$$\frac{\text{Total Variable Cost}}{\text{Number of Guests}} = \text{VC/Guest}$$

- Good managers seek to decrease their fixed costs to their lowest practical levels while still satisfying the needs of the business and its customers. Those same good managers, however, know that *increases* in variable costs are usually very good! You would prefer, for example, to have to purchase extra steaks and incur extra variable costs because that would mean you sold more steaks and increased sales!

Mixed Costs

- It is clear that some business costs are fixed and that some vary with sales volume (variable costs). Still other types of cost contain a mixture of both fixed and variable characteristics. Costs of this type are known as semi-fixed, semi-variable or **mixed costs**.

- In order to fully understand mixed costs, it is helpful to see how the variable and mixed portions are depicted on a mixed cost graph (see Figure 9.3). The x axis represents sales volume and the y axis represents costs in terms of dollars. The total mixed cost line is a combination of fixed and variable costs. The mixed cost line starts at the point where fixed costs meet the y axis. Fixed costs remain the same regardless of sales volume, and thus, must be paid even if no sales occur. The variable cost line, then, sits on top of the fixed cost line, and each guest served or each room sold generates a portion of the variable cost (VC/guest).
- In a hotel, the cost associated with a telephone system is an excellent example of a mixed cost. The hotel will pay a monthly fee for the purchase price repayment, or lease, of the actual phone system. This represents a fixed cost because it will be the same amount whether the occupancy percentage in the hotel is very low or very high. Increased occupancy, however, is likely to result in increased telephone usage by guests. A hotel's local and long distance bill will increase as additional hotel guests result in additional telephone calls made.

Go Figure!

The best way to understand a mixed cost is to understand the mixed cost formula:

> **Fixed Cost + Variable Cost = Total Mixed Cost**
>
> *or*
>
> **Fixed Cost + (Variable Cost per Guest x Number of Guests) = Total Mixed Cost**

Separating Mixed Costs into Variable and Fixed Components

- A mixed cost can be divided into its fixed and variable components in order for management to effectively control the variable cost portion. It is this variable cost that is the most controllable in the short-term.
- Several methods can be used to split mixed costs into their fixed and variable components. The most common methods are high/low, scatter diagrams, and regression analysis. Although regression analysis and scatter diagrams provide more precise results, the high/low method is easier to calculate and gives you a good estimate of the variable and fixed components of a mixed cost.

Go Figure!

The high/low method uses the three following steps:

1. **Determine variable cost per guest for the mixed cost.**

   Choose a high volume month and a low volume month that represents *normal* operations. Then, use the following formula to separate out variable cost per guest for the mixed cost:

> **High Cost − Low Cost**
> **High # of Guests − Low # of Guests** = Variable Cost per Guest (VC/Guest)

2. **Determine total variable costs for the mixed cost.**

    Multiply variable cost per guest by either the high or low volume (number of guests):

    > **VC/Guest x Number of Guests = Total Variable Cost**

3. **Determine the fixed costs portion of the mixed cost.**

    Subtract total variable cost from the mixed cost (at the high volume or low volume you chose in step) to determine the fixed cost portion as follows.

    > **Mixed Cost − Total Variable Cost = Fixed Cost**

Mixed expense can be shown with its variable and fixed components as follows:

> **Fixed Cost + Variable Cost = Total Mixed Cost**
> *or*
> **Fixed Cost + (Variable Cost per Guest x Number of Guests) = Total Mixed Cost**

- In order to determine his variable costs and fixed costs components for a restaurant, a manager can use the following steps.
    1. **Identify all costs as being variable, fixed, or mixed.**
    2. **Determine variable cost per guest for *each* variable cost.**
    3. **Determine *each* fixed cost.** (The best way to identify fixed costs is that they are the same each month.)
    4. **Determine the variable cost and fixed cost portions of *each* mixed cost.** (Use the high/low method to separate mixed costs into their variable and fixed components.)
- For an illustration of variable cost per guest and fixed costs, see Figure 9.5.

Go Figure!

Total costs are mixed costs, and thus can be treated as such. Therefore, by substituting Total Cost for Total Mixed Cost in the Total Mixed Cost formula, Total Costs can be calculated as follows:

> **Fixed Costs + Variable Costs = Total Costs**
> *or*
> **Fixed Costs + (Variable Cost per Guest x Number of Guests) = Total Costs**

- Total fixed costs and total variable costs per guest are the same for all levels of number of guests served. This is because the total cost equation represents a straight line as shown in Figure 9.6.

Go Figure!

As you may remember from high school algebra, the equation for a line is y = a + bx. The equation for a line applies to the Total Cost line, where "a" is the y intercept (fixed costs), "b" is the slope of the line (VC/Guest), "x" is the independent variable (Number of Guests or Sales Volume), and "y" is the dependent variable (Total Cost). The *total cost equation* can be summarized as follows:

> **Total Cost Equation**
> **y = a + bx**
>
> *or*
>
> **Total Costs = Fixed Costs + (Variable Cost per Guest x Number of Guests)**

- Effective managers know they should not categorize fixed, variable, or mixed costs in terms of being either "good" or "bad". Some costs are, by their very nature, related to sales volume. Others are not. The goal of management is not to reduce, but to *increase* total variable costs in direct relation to increases in total sales volume.

Step Costs

- A **step cost** is a cost that increases as a range of activity increases or as a capacity limit is reached. That is, instead of increasing in a linear fashion like variable costs as seen in Figure 9.2, step cost increases look more like a staircase (hence the name "step" costs).

- It is easy to understand step costs. If one well-trained server can effectively provide service for a range of 1 to 30 of a restaurant's guests, and 40 guests are anticipated, a second server must be scheduled. In reality, a 1/3 server would be sufficient, but servers come in "ones". Each additional server added increases the restaurants costs in a non-liner (step-like) fashion. (See Figure 9.7.)

Direct and Indirect (Overhead) Costs

- When a cost can be directly attributed to a specific area or profit center within a business, it is known as a **direct cost.** Direct costs usually (but not always) increase with increases in sales volume.
- An **indirect cost** is one that is not easily assigned to a specific operating unit or department.

- In the hotel industry, indirect costs are more often known as undistributed expenses and non-operating expenses (see Chapter 3). Typically, undistributed expenses include administrative and general, information systems, human resources, security, franchise fees, transportation, marketing, property operations and maintenance, and utility expenses. Other indirect costs include non-operating expenses such as rent and other facility occupation costs, property taxes, insurance, depreciation, amortization, interest, and income taxes.
- Indirect costs are also known as **overhead costs**. When there is more than one profit center, management typically will use a **cost allocation** system to assign portions of the overhead costs among the various centers.

Go Figure!

One approach to cost allocation could be to assign each profit center an equal amount of the business's overhead. If such an approach were used, the allocation would be computed as:

$$\frac{\text{Total Overhead}}{\text{Number of Profit Centers}} = \text{Overhead Allocation per Profit Center}$$

- Another approach would be to assign overhead costs on the basis of the size of each profit center (see Figure 9.10).
- Yet another approach that could be taken to allocate overhead costs, and one that is commonly used by restaurant and hotel companies, is based on sales revenue achieved by the profit center (see Figure 9.11).
- For each different allocation approach utilized, the resulting charges to the individual profit centers are also different.

Controllable and Non-Controllable Costs
- **Controllable costs** are those costs over which a manger has primary control, and **non-controllable costs** are those costs which a manager cannot control in the short-term.
- In most businesses, managers will only be held responsible for the profits remaining after subtracting the expenses they can directly control. In a hotel, examples of controllable costs are operating department expenses and undistributed operating expenses.
- To illustrate, consider the case of Steve, the operator of a neighborhood tavern/sandwich shop. Advertising expense is under Steve's direct control and, thus, would be considered a controllable cost. Some of his expenses, however, are not under his control. The state in which Steve operates charges a tax on all alcoholic beverage sales. The alcoholic beverage tax would be considered a non-controllable cost, that is, a cost beyond Steve's immediate control.
- Experienced managers focus their attention on managing controllable rather than non-controllable costs.

Other Cost Types

- There are additional classifications that managerial accountants often find helpful. Some of these may already be familiar to you. Considering these costs can help managers make better decisions about the operation of their businesses.

Joint Costs

- Closely related to overhead and cost allocation issues is the concept of a **joint cost**. A joint cost is one that should be allocated to two (or more) departments or profit centers.
- Most direct costs are not joint costs, while many indirect costs are considered joint costs.

Incremental Costs

- **Incremental costs** can best be understood as the increased cost of "each additional unit", or even more simply, the cost of "one more".
- Consider the costs incurred by a hotel to sell a single sleeping room to a single traveler. Assume that the managers of a hotel knew that the cost of providing this single sleeping room to a single traveler was $40.00. The direct question related to incremental costs is this, "How much more does it cost to sell the same sleeping room if it is occupied by two guests, rather than one?"

Standard Costs

- The best hospitality managers want to know what their costs *should be*. Bear in mind that management's primary responsibility is not to eliminate costs; it is to incur costs appropriate for the quality of products and services delivered to guests.
- **Standard costs** are defined as the costs that *should* be incurred given a specific level of volume.
- Standard costs can be established for nearly all business expenses from insurance premiums to plate garnishes. For those with experience in the food service industry, understanding standard costs is easy because they already understand standardized recipes. Just as a standardized recipe seeks to describe exactly how a dish should be cooked and served, a standardized cost seeks to describe how much it should cost to prepare and serve the dish. If the variation from the standard cost is significant, it should be of concern to management.

Sunk Costs

- A **sunk cost** is one that has already been incurred and whose amount cannot now be altered. Because it relates to a past decision, information about a sunk cost must actually be disregarded when considering a future decision. Sunk costs are most often identified and considered when making decisions about the replacement or acquisition of assets.

Opportunity Costs

- An **opportunity cost** is the cost of foregoing the next best alternative when making a decision. For example, suppose you have two choices, A and B, both having potential benefits or returns for you. If you choose A, then you lose the potential benefits from choosing B (opportunity cost).
- Opportunity costs are often computed when organizations must choose between several similar, but not completely equal, courses of action.

Cost/Volume/Profit Analysis

- Experienced managerial accountants know that, for most hospitality businesses, some accounting periods are simply more profitable than others. This is so because most businesses experience "busy" periods and "slow" periods.
- For most hospitality businesses, costs as a percentage of sales are reduced when sales are high, and increase when sales volume is lower. The result, in most cases, is greater profits during high volume periods and lesser profits in lower volume periods.
- This relationship between volume, costs, and profits is shown graphically in Figure 9.12, where the x (horizontal) axis represents sales volume. In a restaurant, this is the number of covers (guests) served, or dollar volume of sales. In a hotel, it is the number of rooms sold. The y (vertical) axis represents the costs associated with generating the sales.
- The Total Revenues line starts at 0 because if no guests are served or no rooms are sold, no revenue dollars are generated. The Total Costs line starts farther up the y axis because fixed costs are incurred even if no covers are sold.
- The point at which the two lines cross is called the **break-even point**. At the break-even point, operational expenses are exactly equal to sales revenue. Stated in another way, when sales volume in a business equals the sum of its total fixed and variable costs, its break-even point has been reached. Below the break-even point, costs are higher than revenues, so losses occur. Above the break-even point, revenues exceed the sum of the fixed and variable costs required to make the sales, so profits are generated.

Computation of Cost/Volume/Profit Analysis

- By determining the break-even point, the manager is answering the question, "How much sales volume must I generate before I begin to make a profit?" Beyond the break-even point, the manager will want to answer another question, "How much sales dollars and volume must I generate to make my *target* profit level?"
- A **cost/volume/profit (CVP) analysis** predicts the sales dollars and volume required to achieve a break-even point or desired profit based on known costs.
- Before a CVP analysis can be conducted, a contribution margin income statement must be developed. A **contribution margin income statement** shows P&L items

in terms of sales, variable costs, contribution margin, fixed costs, and profit. The **contribution margin** for the overall operation is defined as the dollar amount, after subtracting variable costs from total sales, that *contributes* to covering fixed costs and providing for a profit. For an illustration of a contribution margin income statement, see Figure 9.13.

Go Figure!

The contribution margin calculation is as follows:

$$\boxed{\text{Total Sales} - \text{Variable Costs} = \text{Contribution Margin}}$$

- The contribution margin income statement can also be viewed in terms of per guest and percentage sales, variable costs, and contribution margin as shown in Figure 9.14.
- Notice the boxed information in Figure 9.14 includes per guest and percent calculations. These include selling price, (SP), variable costs (VC), and contribution margin (CM). See *Go Figure!* in the text for the steps to follow in calculating these numbers.
- Note also that fixed costs are not calculated as per unit or as a percentage of sales. This is because fixed costs do not vary with sales volume increases.

Go Figure!

To determine the dollar sales required to break-even, the following formula is used:

$$\boxed{\frac{\text{Fixed Costs}}{\text{Contribution Margin \%}} = \text{Break-Even Point in Sales Dollars}}$$

The following formula is used to determine the number of guests that must be served in order to break-even:

$$\boxed{\frac{\text{Fixed Costs}}{\text{Contribution Margin per Guest}} = \text{Break-Even Point in Guests Served}}$$

To determine sales dollars and covers needed to achieve his after-tax profit goal, the following formula is used:

$$\boxed{\frac{\text{Fixed Costs} + \text{Before-Tax Profit}}{\text{Contribution Margin \%}} = \text{Sales Dollars to Achieve Desired After-Tax Profit}}$$

The preceding formula calls for *before-tax profit*. To convert after-tax profit to before-tax profit, the following must be calculated:

$$\frac{\text{After-Tax Profit}}{1 - \text{Tax Rate}} = \text{Before-Tax Profit}$$

Sales dollars to achieve desired after-tax profit is computed as follows:

$$\frac{\text{Fixed Costs} + \text{Before-Tax Profit}}{\text{Contribution Margin \%}} = \text{Sales Dollars to Achieve Desired After-Tax Profit}$$

The following formula is used to calculate the number of guests that must be served in order to make the desired after-tax profit:

$$\frac{\text{Fixed Costs} + \text{Before-Tax Profit}}{\text{Contribution Margin per Guest}} = \text{Number of Guests to Achieve Desired After-Tax Profit}$$

- You *must always* round the number of guests *up* because 1) a guest (person) does not exist as a fraction, and 2) it is better to slightly overstate the number of guests to achieve break-even or desired profits than to understate the number and risk a loss or reduce profit. It is better to be safe than sorry!
- Once you round the number of guests up, you should adjust the total sales dollars to reflect this. This difference is minimal and may not warrant adjustment unless an exact sales dollar amount is required based on number of guests.
- When calculating sales and guests (or rooms) to achieve break-even and desired after-tax profits, you can easily remember which formulas to use if you know the following:
  1. Contribution margin % is used to calculate sales *dollars*.
  2. Contribution margin per *guest (or room)* is used to calculate sales volume in *guests (or rooms)*.
- Once you fully understand these CVP analysis concepts, you can predict any sales level for break-even or after-tax profits based on your selling price, fixed costs, variable costs, and contribution margin. You can also make changes in your selling prices and costs to improve your ability to break-even and achieve desired profit levels.

Margin of Safety

- **Margin of safety** shows how close a projected amount of sales will be to break-even, and thus, how close an operation will be to incurring a loss. Margin of safety calculates the difference between projected sales and break-even sales. This calculation is illustrated in Figure 9.15.
- Once the margin of safety for the month is calculated, it can be divided by the number of days in the month to show the margin of safety per day.
- You *must always* round the number of guests for margin of safety *down* because 1) a guest (person) does not exist as a fraction, and 2) it is better to slightly

understate the number of guests as a safety margin than to overstate the number and thus, your safety net. It is better to be safe than sorry!
- Once you round the number of guests down, you should adjust the margin of safety dollars to reflect this. This difference is minimal and may not warrant adjustment unless an exact sales dollar amount is required based on number of guests.

Minimum Sales Point

- A **minimum sales point (MSP)** is defined as the dollar sales volume required to justify staying open for a given period of time. Hotel managers do not typically compute MSPs because, except for seasonal hotels, their operations must be open every day of the week even if only a very few guests are staying in the hotel on any specific day or during any specific time period.
- The information necessary to compute a MSP is as follows:
    - Food cost %
    - Minimum payroll cost for the time period
    - Variable cost %
- Fixed costs are eliminated from the calculation because, even if the volume of sales equals zero, fixed costs still exist and must be paid.

Go Figure!

When a foodservice manager calculates MSP, food cost % + variable cost % is called the **minimum operating cost,** as shown here:

$$\frac{\text{Minimum Labor Cost}}{1 - \text{Minimum Operating Cost}} = MSP$$

or

$$\frac{\text{Minimum Labor Cost}}{1 - (\text{Food Cost \%} + \text{Variable Cost \%})} = MSP$$

Key Terms & Concepts Review
Match the key terms with their correct definitions.

1. Activity based costing _____ a. The point that operational expenses are exactly equal to sales revenue.

2. Activity based management _____ b. The total of food cost percentage plus variable cost percentage found in the denominator of the calculation for minimum sales point.

3. Fixed cost _____ c. Also referred to as Indirect cost.

4. Variable cost _____
5. Mixed cost _____
6. Step cost _____
7. Direct cost _____
8. Indirect cost _____
9. Overhead cost _____
10. Cost allocation _____
11. Controllable cost _____
12. Non-controllable cost _____
13. Joint cost _____
14. Incremental cost _____
15. Standard cost _____
16. Sunk cost _____
17. Opportunity cost _____
18. Break even point _____
19. Cost/volume/profit (CVP) analysis _____
20. Contribution margin income statement _____
21. Contribution margin _____
22. Margin of safety _____

d. A cost that should be allocated to two (or more) departments or profit centers.

e. A cost that is not easily assigned to a specific operating unit or department.

f. The cost of foregoing the next best alternative when making a decision.

g. The dollar amount, after subtracting variable costs from total sales, that contributes to covering fixed costs and providing for a profit.

h. A cost which a manager cannot control in the short-term.

i. A cost that contains a mixture of both fixed and variable cost characteristics.

j. An income statement that shows items in terms of sales, variable costs, contribution margin, fixed costs, and profit.

k. Using activity based costing to examine expenses.

l. A process that assigns each employee's time to different activities performed and then determines the cost spent on each activity as a percentage of each worker's time and pay.

m. The dollar sales volume required to justify staying open for a given period of time.

n. A cost that can be directly attributed to a specific area or profit center within a business.

o. The approach that predicts the sales dollars and volume required to achieve a break-even point or desired profit based on known costs.

p. Cost that remains constant despite increases or decreases in sales volume.

q. A system used by management to assign portions of overhead costs among various profit centers.

r. A cost that increases as a range of activity increases or as a capacity limit is reached.

s. A cost that increases as sales volume increases and decreases as sales volume decreases.

t. The amount that shows how close a projected amount of sales will be to break even, and thus, how close an operation will be to incurring a loss.

u. A cost over which a manager has primary control.

v. The cost that should be incurred given a specific level of volume.

23. Minimum sales point (MSP) _____
24. Minimum operating cost _____

w. A cost that has already been incurred and whose amount cannot now be altered.

x. The increased cost of each additional unit.

Discussion Questions

1. Identify ten ways business costs can be classified.

2. Explain the difference between fixed, variable, and mixed costs.

3. List the three steps in a high/low analysis of costs.

4. Define cost/volume/profit analysis.

Quiz Yourself

Choose the letter of the best answer to the questions listed below.

1. What formula would be used to show that the lower a business's costs, the greater are its profits?
   a. Profit = Costs / Revenue
   b. Profit = Revenue / Costs
   c. Revenue = Profit + Costs
   d. Revenue = Costs – Profit

2. What formula would be used to compute variable cost per guest?
   a. Total variable costs / number of guests
   b. Number of guests / total variable costs
   c. Variable costs – fixed costs
   d. Fixed costs – variable costs

3. Which of the following formulas could be used to calculate mixed costs?
   a. Fixed cost / (variable cost per guest x number of guests)
   b. Fixed cost + (variable cost per guest x number of guests)
   c. (Variable cost per guest x number of guests) + total variable cost
   d. (Variable cost per guest x number of guests) / total variable cost

4. What formula would be used to determine the variable cost per guest?
   a. (High cost – Low cost) / (High # of guests – Low # of guests)
   b. (High cost – Low cost) / (High # of guests / Low # of guests)
   c. (High # of guests – Low # of guests) / (High cost – Low cost)
   d. (High # of guests / Low # of guests) / (High cost – Low cost)

Questions 5, 6 and 7 are based on the following historical data for Chin's Chinese Cuisine for October, November, and December:

	October	November	December
NUMBER OF GUESTS	**10,000**	**17,000**	**21,000**
Food Cost & Beverage Cost	39,000	66,300	81,900
Salaries, Wages, & Employee Benefits	29,085	32,725	34,805
Direct Operating Expenses	14,675	18,735	21,055
Music and Entertainment	1,070	1,070	1,070
Administrative and General	5,570	5,570	5,570
Occupancy, Depreciation, & Interest	20,600	20,600	20,600
Total Costs	**110,000**	**145,000**	**165,000**

5. Using Chin's expense for salaries, wages, and employee benefits, what is the variable cost per guest for this expense?
   e. $3.16
   f. $2.64
   g. $1.65
   h. $0.52

6. What is the total variable cost for the salaries, wages, and employee benefits expense, using the low volume of guests?
   a. $ 5,720
   b. $10,920
   c. $ 5,200
   d. $31,600

7. What is the fixed cost portion of the salaries, wages, and employee benefits expense?
   a. $23,365
   b. $29,602
   c. $23,885
   d. $16,574

8. What formula would be used to calculate contribution margin?
    a. Total sales – variable costs
    b. Total sales + variable costs
    c. Fixed costs + variable costs
    d. Total sales – (fixed costs + variable costs)

9. Assume that Chin's restaurant has an average sale price of $8.00 and a variable cost of $3.20 per unit. If fixed costs are $38,400, how many guests does Chin need in order to break even?
    a. 11,875
    b. 8,000
    c. 4,800
    d. 12,000

10. Assume that Chin's restaurant has an average sale price of $8.00 and a variable cost of $3.20 per unit. If fixed costs are $38,400, and the desired after-tax profit is $6,000, how many guests are needed to achieve this desired profit? The tax rate is 40%.
    a. 10,083
    b. 9,250
    c. 15,125
    d. 10,084

Part IV: Accounting Information for Planning

Chapter 10

Forecasting in the Hospitality Industry

Highlights

- * The Importance of Accurate Forecasts
- * Forecast Methodology
    - Forecasting Restaurant Revenues
    - Forecasting Hotel Revenues
- * Utilizing Trend Lines in Forecasting
- * Key Terms and Concepts Review
- * Discussion Questions
- * Quiz Yourself

Study Notes

The Importance of Accurate Forecasts

- One of the first questions restaurateurs and hoteliers must ask themselves is very simple: "How many guests will we serve today? This week? This year?" The answers to questions such as these are critical, since these guests will provide the revenue from which basic operating expenses will be paid.
- Labor required to serve the guests is also determined based on the manager's "best guess" of the projected number of customers to be served and what these guests will buy.
- Forecasts of future revenues are normally based on a careful recording of previous sales, since what has happened in the past in an operation is usually the best predictor of what will happen in that same operation in the future.
- Finally, operating, cash, and capital budgets (see Chapter 11) cannot be prepared unless an operator knows the amount of revenue upon which these budgets should be based.
- In its simplest case, *sales* are the dollar amount of revenue collected during some predetermined time period. When used in this manner, sales and revenue are interchangeable terms. A distinction is made in the hospitality industry between sales (revenue) and sales volume, which is the number of units sold.
- With accurate sales records, a sales history can be developed for each foodservice outlet you operate, and better decisions will be reached with regard to planning for each unit's operation. In a similar manner, knowing the number of hotel rooms

to be sold allows hoteliers to staff appropriate numbers of room attendants, laundry workers, front desk staff, and food and beverage employees. Because operating costs are typically well-known, the identification of sales levels naturally leads to improved identification of estimated expenses.
- Managers utilizing forecasts and forecast data understand some basic truths about forecasts. These include:
  1. **Forecasts involve the future.** Managers can make long-term forecasts and simply modify them as more precise short-term forecast data become available.
  2. **Forecasts rely on historical data.** In some cases, historical information may be missing, or the accuracy of historical data that does exist may sometimes be questionable. In other cases, unusual events may serve to make historical data misleading. For these reasons, managers utilize historical data, but then carefully add their own estimates and predictions about how that data will vary or be repeated in the future.
  3. **Forecasts are best utilized as a "guide."** Forecasts should be accurate, be based upon the best information available, and be free from unwarranted influence or bias. When forecasts have these characteristics, they can indeed serve as useful managerial guides.

Forecast Methodology

- If only historical data was used to predict future data, forecasting (at least for operations that are already open) would seem to be simple. In fact, in most cases, variations from revenue forecasts are likely to occur.
- When a variation does occur, experienced managers know that some of it can be predicted. Assume that a restaurant has been, for the past several months, experiencing a 10% increase in sales this year when compared to the same period last year. This **trend**, or directional movement of data over time, of increased sales may be very likely to continue
- Several types of trends may occur that can help a hospitality manager forecast revenues. A **seasonal trend**, or a data pattern change due to seasonal fluctuations, can be fairly accurately predicted because it will happen every year. **Cyclical trends** tend to be longer than a period of one year and might occur due to a product's life cycle, such as the downturn of revenues after the "new" wears off of a trendy concept.
- Finally there can simply be **random variation**. This variation appears to occur on a totally unpredictable basis. Upon closer examination however, some random events can be identified. The ultimate goal you should set for yourself as a professional hospitality manager responsible for forecasting sales revenues, expenses, or both is to better understand, and thus actually be able to predict, as much of this random variation as possible.

Forecasting Restaurant Revenues

- For operating restaurants, accurate sales histories are essential for forecasting future sales. A **sales history** is the systematic recording of all sales achieved during a predetermined time period, and is the foundation of an accurate sales forecast. When you predict the number of guests you will serve and the revenues they will generate in a given future time period, you have created a **sales forecast**.
- The simplest type of sales history records revenue only. The sales history format shown in Figure 10.1 is typical for recording sales revenue on a weekly basis.
- Some hospitality managers do not have the ability to consider sales in terms of revenue generated (see Figure 10.2 for an example of the type of sales history used when no cash sales are typically reported). This non-revenue sales history approach is often used where knowledge of the number of actual guests served during a given period is critical for planning purposes.
- For most restaurant operations, sales revenue consists of the number of guests served as well as how much each of those guests spend. Sales histories can be developed to track the number of guests served as well as to compute average sales per guest, a term commonly known as check average (see Chapter 6). Recall that the formula for average sales per guest is:

$$\frac{\text{Total Sales}}{\text{Number of Guests Served}} = \text{Average Sales per Guest (Check Average)}$$

- Maintaining histories of average sales per guest is valuable if they are recorded as weighted averages. The reason why can be seen easily by analyzing the data in Figure 10.3.
- The use of a **simple average**, or the value arrived at by adding the quantities in a series and dividing that sum by the number of items in the series, to calculate average sales per guest *would not be correct*.
- In fact, the **weighted average sales per guest**, or the value arrived at by dividing the total amount guests spend by the total number of guests served, should be used.
- In most cases, sales histories should be kept for a period of at least two years. This allows managers to have a good sense of what has happened in their business in the recent past as well as the most recent time periods.

Forecasting Future Revenues

- To learn how managers use sales histories to forecast future sales, consider the sales history shown in figure 10.4.

Go Figure!

The dollar variance for sales for the period is computed as follows:

$$\text{Sales This Year} - \text{Sales Last Year} = \text{Variance}$$

The percentage variance for the period is computed as follows:

$$\frac{\text{Variance}}{\text{Sales Last Year}} = \text{Percentage Variance}$$

Another way to compute the percentage variance is to use a math shortcut, as follows:

$$\frac{\text{Sales This Year}}{\text{Sales Last Year}} - 1 = \text{Percentage Variance}$$

It is important to understand that sales histories do not record the reasons for increases or decreases in revenue. Knowing those is the job of the operation's manager. See Figure 10.5 for an illustration of how a planning sheet for the next period could be developed.

Go Figure!

For the next period total, the sales forecast is calculated as follows:

$$\text{Sales Last Year} + (\text{Sales Last Year} \times \text{\% Increase Estimate}) = \text{Sales Forecast}$$

An alternative way to compute the sales forecast is to use a math shortcut, as follows:

$$\text{Sales Last Year} \times (1 + \text{\% Increase Estimate}) = \text{Sales Forecast}$$

Forecasting Future Guest Counts

- Using the same techniques employed in estimating increases in sales, the noncash operator or any manager interested in guest counts can estimate increases or decreases in the number of guests served (see Figure 10.6).
- See Figure 10.7 for an illustration of how a planning sheet for predicting the guest count for the next period could be developed.

Go Figure!

For the next period total, guest count forecast is calculated as follows:

$$\text{Guest Count Last Year} + (\text{Guest Count Last Year} \times \text{\% Increase Estimate}) = \text{Guest Count Forecast}$$

This process can be simplified by using a math shortcut, as follows:

> **Guest Count Last Year X (1.00 + % Increase Estimate) = Guest Count Forecast**

Forecasting Future Average Sales per Guest

- Average sales per guest (check average) is simply the amount of money an average guest spends during a visit. The same formula is used to forecast average sales per guest as was used in forecasting total revenue and guest counts. Therefore, using data taken from an operation's sales history, the following formula is employed:

> **Last Year's Average Sales per Guest + Estimated Increase in Sales per Guest = Sales per Guest Forecast**

- Alternatively, you can compute average sales per guest using the data collected from sales forecasts divided by the data collected from guest count forecasts. Figure 10.8 illustrates how the data from Figures 10.5 and 10.7 can be combined to compute a sales per guest forecast.

Go Figure!

For the second period total, the average sales per guest forecast is determined as follows:

> $$\frac{\text{Sales Forecast}}{\text{Guest Count Forecast}} = \text{Average Sales per Guest Forecast}$$

- Sales histories, regardless of how well they have been developed and maintained, are not, when used alone, sufficient to accurately predict future sales. The operation's managers must have knowledge of potential price changes, new competitors, facility renovations, and improved selling programs to name just a few of the many factors that must be considered when predicting future revenues.

Forecasting Hotel Revenues

- Accurate and useful forecasts are even more important to hotels than they are to restaurants. This is so because, unlike restaurateurs, hoteliers are most often held accountable not only for controlling costs, but also for the short-term management of revenues via yield management and other RevPAR maximization techniques. In addition, hotel room rates (unlike most restaurant menu prices) are likely to be adjusted daily, weekly, and monthly based upon a hotel's forecast of future demand for its hotel rooms.
- Forecasts that are 100% accurate are rare. Forecasts that are consistently or significantly in error, however, will ultimately result in significant financial or

operational difficulty for a hotel. This is true whether the forecasts are too high or too low. Forecasts that are implausibly too high:
- Cause unrealistic expectations by the hotel's owners
- Increase feelings of frustration by affected staff when forecasted volume levels are not attained
- Produce budgeting/ spending errors by overstating anticipated revenues
- Result in impractical and overly aggressive room rate determinations. When forecasts are excessively high, room rates may be set too high (see Chapter 8).
- Alternatively, forecasts which are consistently and unrealistically too low:
  - Lead management to believe it is actually performing at higher levels of room sales
  - Undermine the credibility of the forecaster(s) because of the suspicion that actual forecast variation is due to **low-balling** (intentionally underestimating) the forecasts
  - Result in impractical and under-aggressive room rate determinations (rates too *low)*
- With experience, those responsible for forecasting demand for a hotel's rooms can produce room forecasts that will be within 1% to 5% of actual hotel room revenues.
- Accurate demand forecasts are not an end, but rather, as you learned in Chapter 8, a means of improving the effectiveness of establishing room rates. **Demand forecasts** predict periods in which demand will be generally high or low based on expected fluctuations in occupancy. Types of hotel forecasts can be found in Figure 10.9.
- While restaurateurs rely primarily on historical (and in some cases predictive future data), to estimate revenues and guests to be served, hoteliers must rely on a combination of historical, current, and future data to accurately forecast and manage room demand.
- Historical data, of course, refers to events that have happened in the past. Current data is related to events that are entered into a hotel's property management system (PMS) but have yet to occur. Accurate current data reflects confirmed (but as of yet not historic) information about the demand for hotel rooms. Future data is that information which is related to events that have yet to occur and are not currently be found in the hotel's PMS database.
- Hoteliers create occupancy forecasts in a manner similar to how they compute actual occupancy percentage. That is:

$$\frac{\text{Rooms Sold}}{\text{Rooms Available for Sale}} = \text{Actual Occupancy \%}$$

becomes

$$\frac{\text{Rooms Forecasted to be Sold}}{\text{Rooms Available for Sale}} = \text{Occupancy Forecast \%}$$

- Increasingly, PMSs are designed to include forecasting programs or components. While PMS forecasting modules may indeed "guide" them, local property

managers themselves will ultimately make the best occupancy forecasts for their own hotels because they:
- Understand the unique property features affecting demand for their own hotels
- Know about special city-wide events in the area that affect demand
- Have an understanding of the demand for competitive hotel properties in the area
- Can factor in the opening or closing of competitive hotels in the area
- Can include factors such as weather, road construction, and seasonality in their demand assessments
- Can adjust forecasts very quickly when faced with significant demand-affecting events (i.e., power outages and airport or highway closings)

Historical Data

- One of the best ways for existing hotels to predict future room demand is by examining historical demand. Managers combine their own skills and experience with relevant historical (and other) data when creating usable demand forecasts. In order to create these forecasts, you must first understand the following terms:
    - **Stayover**: A guest that is not scheduled to check out of the hotel on the day his or her room status is assessed. That is, the guest will be staying and using the room(s) for at least one more day.
    - **No-show**: A guest who makes a room reservation but fails to cancel the reservation (or arrive at the hotel) on the date of the reservation.
    - **Early Departure**: A guest who checks out of the hotel *before* his or her originally scheduled check-out date.
    - **Overstay**: A guest who checks out of the hotel *after* his or her originally scheduled check-out date.
- Figure 10.10 shows the method used to compute a single day's occupancy forecast for a 300 room hotel. In an actual hotel setting, room usage and availability would be forecast by individual room type, as well as for the total number of hotel rooms available. The procedures and steps are the same when forecasting room type availability and/or total room availability.
- The number of hotel rooms available, the number of out of order rooms, the number of stayovers, and the number of reservations currently booked are all data that resides in the PMS. Data for the three forecast adjustments (no-shows, early departures, and overstays), however, describe events that will occur in the future, and thus "real" data on them does not exist. These numbers must be forecast by managers after carefully tracking the hotel's historical data related to them.

Using Current and Future Data

- Historical data in the PMS is very valuable because room demand often follows fairly predictable patterns. However, the use of historical data alone is, most often, a very poor way in which to forecast room demand. Current and future data must also be assessed.

- In the situation presented in Figure 10.11, estimated demand based on the historical data would be far short of the demand indicated by the current data. **On-the-books** is the term hoteliers use to describe current data and it is used in reference to guest reservations. The term originated in the days when hotel reservation data was stored in a bound reservations book.
- Future data is the final type of information needed to assist hoteliers in accurately forecasting demand. In fact, most hoteliers agree that a manager's ability to accurately assess this information is the most critical determiner of an accurate demand forecast.
- As can be seen from the examples given in the text, if hoteliers are to make accurate forecasts and properly price their rooms, historical, current, and future data must all be carefully considered. This is so because occupancy forecasting is not simply a matter of identifying the number of hotel rooms that may be sold, but rather it is a multifaceted process that consists of four essential activities that include:
  - Generating the demand forecast
  - Establishing a room rate strategy
  - Monitoring reservation activity reports
  - Modifying room rate pricing strategies (if warranted)
- For hoteliers (unlike restaurateurs), pricing decisions naturally follow forecast development. Thus, accurate demand forecasts will profoundly affect a professional hotelier's room pricing decisions.
- Only by creating an accurate forecast can a hotel know when room demand is strong or weak enough to dictate significant changes in pricing strategies and thus affect the procedures and tactics designed to help a hotel achieve its RevPAR and revenue per occupied room (RevPOR) goals.

Utilizing Trend Lines in Forecasting

- Sophisticated business forecasters in many fields have found that the application of advanced mathematical models can often result in forecasts which are much more accurate than those derived from approaches which do not utilize such advanced formulas.
- Fortunately, such formulas exist in somewhat basic forms and are easy for most managers to use. One such popular formula produces a **trend line**, which is a graphical representation of trends in data that you can use to make predictions about the future. For the purpose of this chapter, a trend line will be used to forecast future sales.
- This analysis is also called a regression analysis. A **regression analysis** estimates an activity (dependent variable - forecasted sales in this case) based on other known activities (independent variables - past sales in this case). By using a regression analysis, you can extend a trend line in a chart beyond the actual known data for the purpose of predicting future (unknown) data values.
- Once the annual sales data for several fiscal years has been collected (see Figure 10.12) a line graph can be created using Excel (see Figure 10.13).

- A trend line (future prediction) can be created, using the **baseline data** (known data). All managers creating trend lines must ensure that:
  - There is enough data to show a meaningful trend. Insufficient baseline data will likely skew results.
  - The data is entered into the spreadsheet from earliest (oldest) to most recent (newest).
  - No data is missing. If data is unavailable for a period, an estimate must be entered.
  - All periods are for comparable amounts of time, such as weeks, months, or, as is the case in this example, fiscal years.
- If all of the above items are satisfactory, a trend line can be crated to predict future sales levels using Excel (see Figure 10.14).

Key Terms & Concepts Review

Match the key terms with their correct definitions

1. Trend _____
2. Seasonal trend _____
3. Cyclical trend _____
4. Random variation _____
5. Sales history _____
6. Sales forecast _____
7. Simple average _____
8. Weighted average sales per guest _____
9. Low-balling _____
10. Demand forecast _____
11. Stayover _____

a. An analysis that estimates an activity (dependent variable) based on other known activities (independent variables).

b. A guest who checks out of the hotel before his or her originally scheduled check-out date.

c. The term hoteliers use to describe current data in reference to guest reservations.

d. A guest who checks out of the hotel after his or her originally scheduled check-out date.

e. A data pattern change due to seasonal fluctuations.

f. Known data used to predict trend lines

g. A graphical representation of trends in data that can be used to make predictions about the future.

h. A systematic recording of all sales achieved during a predetermined time period.

i. The directional movement of data over time.

j. A prediction of the number of guests served and the revenues generated in a given future time period.

k. A data pattern that tends to be longer than a period of one year and might occur due to a product's life cycle.

12. No-show _____

13. Early departure _____

14. Overstay _____

15. On-the-books _____

16. Trend line _____

17. Regression analysis _____

18. Baseline data _____

l. The value arrived at by adding the quantities in a series and dividing that sum by the number of items in the series.

m. A guest that is not scheduled to check out of the hotel on the day his or her room status is assessed.

n. The act of intentionally underestimating.

o. A data variation that appears to occur on a totally unpredictable basis.

p. A guest who has a guaranteed reservation and neither cancels the reservation nor shows up at the hotel on their expected date of arrival.

q. The value arrived at by dividing the total amount guests spend by the total number of guests served.

r. A prediction of periods in which demand will be generally high or low based on expected fluctuations in occupancy.

Discussion Questions

1. Identify three basic truths about forecasts.

2. Explain the term "weighted average sales per guest."

3. List the reasons local property managers can make the best forecasts for their hotels.

4. List the four essential activities for forecasting occupancy in a hotel.

Quiz Yourself
Choose the letter of the best answer to the questions listed below.

1. The prediction of the number of guests you will serve and the revenues they will generate in a given time is called:
   a. Sales history
   b. Variance
   c. Sales forecast
   d. Variance percentage

2. In the first quarter of the year, Pete's Place served 4,907 guests in January, 5,234 guests in February, and 4,256 guests in March. Pete's total sales were $57,600 in January, $45,300 in February, and $48,700 in March. What was his weighted check average for the first quarter of the year?
   a. $10.61
   b. $10.53
   c. $11.74
   d. $ 8.65

3. This year, sales at Pete's Place totaled $720,000. What is his variance from last year?
   a. $1,412,000
   b. $ 104
   c. $ 498,240
   d. $ 28,000

4. Given the variance from question 3, what is the percentage variance?
   a. 3.89%
   b. 4.05%
   c. 96.11%
   d. 95.95%

5. Joe's Joint had sales this year of $435,000. With an expected increase of 3.45%, what is his sales forecast for next year?
   a. $450,007.50
   b. $419,992.50
   c. $468,232.80
   d. $854,992.50

6. Joe's Joint had a guest count of 55,769 this year and Joe has calculated an estimated guest count increase of 0.86% for next year. What is his guest count forecast?
   a. 57,093
   b. 56,249
   c. 55,289
   d. 60,565

7. Average sales per guest this year at Joe's Joint was $7.80. Based on his estimated forecasts in sales and guest counts, from questions 5 and 6, what is his average sales per guest forecast for next year?
   a. $7.88
   b. $7.43
   c. $8.00
   d. $8.14

8. Between Pete's Place and Joe's Joint is Harry's Hotel, which has 24 rooms available for sale. Last year the hotel sold an average of 18 rooms per night. What was the occupancy percentage?
   a. 66.7%
   b. 71.2%
   c. 73.5%
   d. 75.0%

9. Based on the sales growth this year, Harry's Hotel forecasts selling an average of 20 rooms per night. What is the occupancy forecast %?
   a. 83.3%
   b. 75.0%
   c. 66.7%
   d. 92.5%

10. Another name for an analysis which produces a trend line is
    a. Scatter analysis
    b. Sales analysis
    c. Regression analysis
    d. Variance analysis

Chapter 11

Budgeting and Internal Controls

Highlights

* Importance of Budgets
* Types of Budgets
* Operations Budget Essentials
* Developing an Operations Budget
* Monitoring an Operations Budget
* Cash Budgeting
* Managing Budgets through Internal Controls
* Key Terms and Concepts Review
* Discussion Questions
* Quiz Yourself

Study Notes

The Importance of Budgets

- Just as the income statement tells a managerial accountant about past performance, the **budget,** or financial plan, is developed to help you achieve your future goals. In effect, the budget tells you what must be done if predetermined profit and cost objectives are to be met.
- Without such a plan, you must guess about how much to spend and how much sales you should anticipate. Effective managers build their budgets, monitor them closely, modify them when necessary, and achieve their desired results.
- While each organization may approach budgeting from its own perspectives and within its own guidelines, a budget is generally produced by:
  1. Establishing realistic financial goals of the organization
  2. Developing a budget (financial plan) to achieve the objectives
  3. Comparing actual operating results with planned results
  4. Taking corrective action, if needed, to modify operational procedures and/or modify the financial plan
- The advantages of preparing and using a budget are summarized in Figure 11.1.
- Budgeting is best done by the entire management team, for it is only through participation in the process that the whole organization will feel compelled to support the budget's implementation.
- In large organizations, a variety of individuals will be involved in the budgeting process. The **Chief Executive Officer (CEO),** the highest ranking officer in charge of the overall management of a company, is ultimately responsible for the

company's financial performance primarily due to the Sarbanes-Oxley Act (see Chapter 1). Because of this, the budgeting process will begin with the CEO establishing financial goals for the company's profitability. From this, a company can develop a strategic plan to meet its mission and objectives, and the budget can serve as a link from the strategic plan to the company's financial goals.
- At the middle levels of large hospitality companies, regional, district, area, and unit managers will be involved in the budgeting process because they will have access to the revenue and cost projections associated with the units for which they are responsible. In many cases, bonuses for these professionals will be directly tied to their ability to accurately forecast revenues and expenses in their units.
- In addition, individual restaurant or hotel owners will want to know what they can expect to earn on their investments. A budget is necessary to project those earnings. In organizations of all sizes, proper budgeting is a process that is of critical importance, and it is equally critical that it is done well.

Types of Budgets

- Experienced managerial accountants know that they may be responsible for helping to prepare not one, but several budgets at the same time. Budgets will need to be created with differing time frames and for very different purposes. "Length" and "purpose" are two of the most common methods of considering the different types of budgets managers prepare.

Length

- One extremely helpful way to consider a budget is by its length or **horizon**. While a budget may be prepared for any time frame desired by a manager, budget lengths are typically considered to be one of three types, as follows:
  1. Long-range budget
  2. Annual budget
  3. Achievement budget

Long-Range Budget

- The **long-range budget** is typically prepared for a period of up to five years. While its detail is not great, it does provide a long-term financial view about where an operation should be going.

Annual Budget

- The **annual budget**, or yearly budget, is the type many hospitality managers think of when the word budget is used.
- An annual budget need not follow a calendar year. In fact, the best time period for an annual budget is the one that makes sense for your own operation. A college foodservice director, for example, would want a budget that covers the time period of a school year.

- An annual budget need not consist of 12, one-month periods. While many operators prefer one-month budgets, some prefer budgets consisting of 13, 28-day periods, while others use quarterly (three-month) or even weekly budgets to plan for revenues and costs throughout the budget year.

Achievement Budget

- The **achievement budget**, or short-range budget, is always of a limited time period, often consisting of a month, a week, or even a day. It most often provides very current operating information and thus, greatly assists in making current operational decisions. For an example of an achievement budget, see *Go Figure!* in the text.

Purpose

- Budgets are also frequently classified based upon their specific purpose. For example, a front office manager in a hotel may want to create a monthly budget for the front desk's cost of labor, a restaurant manager may wish to budget for next week's cost of food, or a club manager may wish to create a budget designed to estimate the club's annual utility costs.
- For most hospitality managers, budgets can be created for use in one of three broad categories, which are:
  - Operations budgets
  - Cash budgets
  - Capital budgets

Operations Budgets

- **Operations budgets** are concerned with planning for the revenues, expenses, and profits associated with operating a business.
- The operations budget is simply management's estimate of all (or any portion of) the income statement.

Cash Budgets

- In Chapter 5 (The Statement of Cash Flows), you learned that cash may be generated or expended by a business's operating activities, investing activities, and financing activities. **Cash budgets** are developed to estimate the actual impact on cash balances that will result from these activities.

Capital Budgets

- As you learned in Chapter 4 (The Balance Sheet), some expenses incurred by a business are not recorded on the income statement. **Capital expenditures** are those expenses associated with the purchase of land, property and equipment, and other fixed assets that are recorded on the balance sheet.

- The **capital budget** is the device used to plan for capital expenditures. Planning (budgeting) for capital expenditures is related to the investment goals of the business's owners, as well as their long-term business plans.
- Figure 11.2 summarizes the three budget types and purposes most commonly utilized.

Operations Budget Essentials

- An operations budget is a forecast of revenue, expenses, and resulting profits for a selected accounting period.
- Before managers can begin to develop an operations budget, they must understand the underpinning essentials required for its creation. Before you begin the process of assembling an operations budget you will need to have and understand the following information:
    - Prior-period operating results (if an existing operation)
    - Assumptions made about the next period's operations
    - Knowledge of the organization's financial objectives

Prior-Period Operating Results

- The task of budgeting becomes somewhat easier when you examine your operation's prior period operating results. The further back, and in more detail, you can track your operation's historical revenues and expenses, the better your budgets are likely to be.
- Historical data should always be considered in conjunction with the most recent data available.

Assumptions about the Next Period's Operations

- Evaluating future conditions and activities are also necessary when developing an operations budget.
- After demand factors have been considered, assumptions regarding revenues and expenses may be made. From these assumptions, projected percentages of increases or decreases in revenues and expenses may be made to develop the operations budget.

Knowledge of the Organization's Financial Objectives

- An operation's financial objectives may consist of a given profit target defined as a percent of revenue or a total dollar amount, as well as specific financial and operational ratios that should be achieved by management (see Chapter 6). The operations budget must incorporate these goals.

Developing an Operations Budget

- The operations budget is a detailed plan which can be expressed by the budget formula as follows:

> **Budgeted Revenue − Budgeted Expense = Budgeted Profit**

- The budgeted profit level a manager seeks can be achieved when the operation realizes the budgeted revenue levels *and* expends only what has been budgeted to generate those revenues. If revenues fall short of forecast and/or if expenses are not reduced to match the shortfall, budgeted profit levels are not likely to be achieved. In a similar manner, if actual revenues exceed forecasted levels, expenses (variable and mixed) will also increase. If the increases are monitored carefully and are not excessive, increased profits should result. If, however, managers allow expenses to exceed the levels actually required by the additional revenue increase, budgeted profits again will not be achieved.

Budgeting Revenues

- Forecasting revenues is critical since all forecasted expenses and profits will be based on revenue forecasts. Revenues should be estimated on a monthly (or weekly) basis and then be combined to create the annual revenue budget, because many hospitality operations have seasonal revenue variations.
- Forecasting revenues is not an exact science. However, it can be made quite accurate if managers implement the following:
- **Review historical records.** Review revenue records from previous years.
- **Consider internal factors affecting revenues.** Any internal initiation or change that management believes will likely impact future revenues should considered in this step.
- **Consider external factors affecting revenues.** There are a variety of external issues that can affect an operation's revenue forecasts.

Go Figure!

After considering internal and external factors affecting revenues, a sales forecast can be computed as follows:

> **Sales Last Year X (1 + % Increase Estimate) = Sales Forecast**

Using historical data and a forecasted increase in selling prices, his forecasted check average will be calculated as follows:

> **Selling Price Last Year X (1 + % Increase Estimate) = Selling Price Forecast**
> **(Check Average)**

Budgeting Expenses

- Managers must budget for each fixed, variable, and mixed cost when addressing the individual **line items,** or expenses, found on the income statement. Fixed costs are simple to forecast because items such as rent, depreciation, and interest typically stay the same from month to month. Variable costs, however, are directly related to the amount of revenue produced by a business. Mixed costs contain both fixed and variable cost components.

Fixed Costs

- In most cases, budgeting fixed costs is quite easy since they remain unchanged regardless of the revenues generated by the restaurant. Of course, any anticipated increases in fixed costs for the year will have to be budgeted in each month.

Variable Costs

- Variable costs increase or decrease as revenue volumes change.
- Variable costs can be forecasted using targeted percentages or costs per unit (rooms or covers). When percentages are used, the sales forecast is simply multiplied by the target cost % to get the forecasted cost.

Go Figure!

A targeted food cost percentage and food sales would yield the following food cost:

Sales Forecast X Targeted Cost % = Forecasted Cost

To forecast costs using per unit (cover) costs, you would use last year's cover costs plus the increase estimates, as follows:

Cost per Cover Last Year X (1 + % Increase Estimate) = Cost per Cover Forecast

Food costs can then be forecasted costs using the following formula:

Cost per Cover Forecast X Forecasted Number of Covers = Forecasted Costs

Mixed Costs

- One of the largest (if not the single largest!) line item costs in a hospitality operation is that of labor. Labor is a mixed cost because it includes hourly wages (variable costs), salaries (fixed costs) and employee benefits (mixed costs).

Accurate budgets used to help control future labor costs can be precisely calculated using a 3-step method.

Step 1: Determine Targeted Labor Dollars to be Spent
- In most cases, the determination of labor costs is tied to the targeted or standard costs an operation seeks to achieve. These standards or goals may be established by considering the historical performance of an operation, by referring to industry segment or company averages, or by considering the profit level targets of the business.

Go Figure!

With a sales forecast and a labor cost percentage standard established, the total amount to be budgeted for labor (salaries, wages, and employee benefits) can be calculated:

> **Sales Forecast X Labor Cost % Standard = Forecasted Labor Cost**

Step 2: Subtract Costs of Payroll Allocations
- **Payroll allocations** consist of those costs associated with, or allocated to, payroll. These non-wage costs must be paid by employers and include items such as payroll taxes as well as voluntary benefit programs that may be offered by the operation.
- The cost of these mandatory (and voluntary) programs can be significant. Depending upon the restaurant or hotel, the cost of voluntary benefit programs that must be reduced from the total dollar amount available for labor include costs such as:
    - Bonuses
    - Health, dental, and vision insurance
    - Life insurance
    - Long-term disability insurance
    - Employee meals
    - Sick leave
    - Paid holidays
    - Paid vacation

Go Figure!

The calculation required to determine the budgeted payroll allocation amount would be:

> **Forecasted Labor Cost X Payroll Allocation % = Budgeted Payroll Allocation**

The amount remaining for use in paying all operational salaries and hourly wages (budgeted payroll) would be computed as:

> **Forecasted Labor Cost − Budgeted Payroll Allocation = Budgeted Payroll**

Step 3: Subtract Salaried (Fixed) Wages to Determine Variable Wages
- Variable payroll is the amount that "varies" with changes in sales volume. The distinction between fixed and variable labor is an important one, since managers may sometimes have little control over their fixed labor costs, while at the same time exerting nearly 100% control over variable labor costs.

Go Figure!

The amount to be budgeted for variable hourly payroll can be calculated as:

Budgeted Payroll – Salaries and Fixed Wages = Budgeted Hourly Payroll

- In most restaurant operations, managers who successfully create an operations budget for food and labor will have accounted for more than 50% of their total costs. In a similar manner, hoteliers that have budgeted for their rooms-related expenses and labor costs are likely to have accounted for more than 50% of their total expenses.
- All other expenses can be budgeted using the same methods for fixed, variable, and mixed costs.
- For an example of an operating budget for a restaurant, see Figure 11.3.
- When every fixed, variable, and mixed cost is included, the result will be an operations budget that:
  1. Is based upon a realistic revenue estimate
  2. Considers all known fixed, variable, and mixed costs
  3. Is intended to achieve the financial goals of the organization
  4. Can be monitored to ensure adherence to the budget's guidelines
  5. May be modified when necessary

Monitoring an Operations Budget

- An operations budget has little value if management does not use it. In general, the operations budget should be regularly monitored in each of the following three areas:
  - Revenues
  - Expenses
  - Profit
- **Revenues**. Effective managers compare their actual revenue to that which they have projected on a regular basis.
- **Expenses**. Effective foodservice managers are careful to monitor operational expenses because costs that are too high or too low may be cause for concern
- **Profit**. In the final analysis, however, budgeted profit must be realized if the operation is to provide adequate returns for owner and investor risk. Management's task is not merely to generate a profit, but rather to generate the

budgeted profit!

Variances

- Managers in all segments of the hospitality industry must ensure operational profitability by analyzing the differences between budgeted (planned for) results and actual operating results. The difference between planned results and actual results is called **variance**.
- Variance may be expressed in either dollar or percentage terms and can be either positive (favorable) or negative (unfavorable). A **favorable variance** occurs when the variance is an improvement on the budget (revenues are higher or expenses are lower). An **unfavorable variance** occurs when actual results do not meet budget expectations (revenues are lower or expenses are higher).

Go Figure!

The variance between actual expense and budgeted expense is calculated as follows:

$$\text{Actual Expense} - \text{Budgeted Expense} = \text{Variance}$$

The variance may be expressed as a dollar amount or as a percentage of the original budget. The computation for percentage variance is:

$$\frac{\text{Variance}}{\text{Budgeted Expense}} = \text{Percentage Variance}$$

Another way to compute the percentage variance is to use a math shortcut, as follows:

$$\frac{\text{Actual Expense}}{\text{Budgeted Expense}} - 1 = \text{Percentage Variance}$$

- The variance is unfavorable if the actual expense is higher than the budgeted expense. A **significant variance** is any difference in dollars or percentage between budgeted and actual operating results that warrants further investigation. Significant variance is an important concept because not all variances should be investigated. For an illustration of significant variances, see *Go Figure!* in the text.
- Managerial accountants must decide what represents a significant variance based on their knowledge of their specific operations as well as their own company policies and procedures. Small percentage differences can be important if they represent large dollar amounts. Similarly, small dollar amounts can be significant if they represent large percentage differences from planned results. Managers can monitor all of these areas using the following operations budget monitoring process:

Operations Budget Monitoring Process	
Step 1	Compare actual results to the operations budget.
Step 2	Identify significant variances.
Step 3	Determine causes of variances.
Step 4	Take corrective action or modify the operations budget.

- In Step 1, the manager studies income statement and operations budget data for a specified accounting period. In Steps 2 and 3, actual operating results are compared to the budget and significant variances, if any, are identified and analyzed. Finally, in Step 4, corrective action is taken to reduce or eliminate unfavorable variances, or if it is appropriate to do so, the budget is modified to reflect new realities faced by the business.
- In most cases, managers will compare their actual results to operations budget results in each of the income statement's three major sections of revenue, expense, and profits (see Figure 11.4).

Revenue Analysis

- Revenues are the first area to be examined when comparing actual results to budgeted results. If revenue falls significantly below projected levels, there will likely be a significant negative impact on profit goals. Variable costs should be less than budgeted. In addition, those fixed and mixed expenses (such as rent and labor) incurred by the operation will represent a larger than budgeted percentage of total revenue. Alternatively, if actual revenues exceed the budget, total variable expenses will increase, while the fixed and mixed expenses incurred by the operation should, if properly managed, represent a smaller than budgeted percentage of total revenue.
- Foodservice operations and hotels that consistently fall short of revenue projections must also evaluate the wisdom and validity of the primary assumptions used to produce the revenue portion of their operations budgets.

Expense Analysis

- Identifying significant variances in expenses is a critical part of the budget monitoring process as many types of operating expenses are controllable expenses (see Chapter 9). The variances that occur can tell managers a great deal about operational efficiencies.
- When specific line item operating expenses vary significantly from the operations budget, those significant variances should be analyzed using the four-step budget "Operations Budget Monitoring Process" presented earlier in this section.

Profit (Net Income) Analysis

- A hospitality operation's actual level of profit measured either in dollars, percentages, or both is simply the most critical number that most hospitality managers must evaluate. The inability of a business to meet its operational revenue budget typically means that the budget was ineffectively developed, internal/external conditions have changed, and/or that the operation's managers were not effective. Regardless of the cause, when revenues do not reach forecasted levels, corrective action is usually needed to prevent even more serious problems including profit erosion..

Budget Modifications

- An operations budget should be regularly reviewed and modified as new and better information replaces the information that was available when the original operations budget was developed.
- An operations budget should never be modified simply to compensate for management inefficiencies. Well-prepared operations budgets are designed to be achieved and managers must do their best to achieve them.
- There are cases, however, when operations budgets simply must be modified or they will lose their ability to assist managers in decision making. The following situations are examples of those that, if unknown and not considered at the time of the original budget development process, may legitimately require managers to contemplate a modification of their existing operations budget:
    o Additions or subtractions from product offerings that materially affect revenue generation (for example, reduced or increased operating hours)
    o The opening of a direct competitor
    o The closing of a direct competitor
    o A significant and long-term or permanent increase or decrease in the price of major cost items
    o Franchisor mandated operating standards changes that directly affect (increase) costs
    o Significant and unanticipated increases in fixed expenses such as mortgage payments, insurance, or property taxes
    o A management (or key employee) change that significantly alters the skill level of the facility's operating team
    o Natural disasters such as floods, hurricanes, or severe weather that significantly affects forecasted revenues
    o Changes in financial statement formats or expense coding policies
    o Changes in the investment return expectations of the operation's owners

Flexible Budgets

- Budget modifications can also be used to help managers better evaluate their performance based on varying levels of sales activity. These budget

modifications result in the creation of flexible budgets. A flexible budget incorporates the assumptions of the original budget, such as fixed costs and target variable costs per unit or variable cost percentages, and then projects these costs based on varying levels of sales volume.
- By developing several budgets at varying levels of covers based on the same assumptions, you could choose the budget with the number of covers closest to actual sales activity (15,000 covers) to evaluate your performance.
- A successful manager will have separated all of the mixed costs into their fixed and variable components using the high/low method described in Chapter 9. Once this is done, it is easy to budget variable costs based on the budgeted sales activity (number of covers) and fixed costs separately. (See Figure 11.5)
- With both the flexible budget and actual results reflecting the same number of covers sold, realistic variances can be calculated. As you learned earlier, the variance is calculated by subtracting the budgeted sales or expenses from the actual sales or expenses. The final outcome is an analysis of favorable (F) or unfavorable (U) results.
- By using flexible budgets, managers are not "stuck" with static budgets that do not accurately reflect their performance.

Cash Budgeting

- Hospitality businesses are like any other business ventures in that their primary objectives are to survive and grow. To stay in business, a company simply must have access to enough cash to meet its obligations. If it does not, it will not survive. Securing sufficient cash to continue operations is not merely an important issue; it is every business's single most important issue.
- "Cash" is considered to consist of all petty cash accounts held in an operation, all cashier bank balances, and all to-be-deposited, as well as previously deposited and unrestricted, cash receipts.
- Generally, more cash is required during high volume periods because payrolls are larger, more cash is tied up holding greater quantities of products in inventory, and in some restaurants as well as most hotels, accounts receivable amounts are higher in busy months
- **Cash budgeting** is the general term used by managerial accountants to identify a variety of cash monitoring and management activities. Effective cash budgeting is such a primary business objective that it often must override other objectives, such as increasing sales volume.
- The concept of total cash receipts (cash in) and cash disbursements (cash out) which occur in a business in a specific accounting period is called **cash flow.** In most businesses, cash flows, and thus changes in cash balances on hand, occur in a predictable cycle.

Cash and Accounts Receivable (AR) Activities

- After the determination has been made that credit terms will be extended, it is important to bill guests according to the policies established by the business. It is equally important to monitor accounts receivable. This is done via the **accounts receivable aging report,** which is used by management to monitor the average length of time money owed to the business has remained uncollected.
- Figure 11.6 shows an example of an accounts receivable aging report.
- As a general rule, as the age of an accounts receivable increases, its likelihood of being collected decreases. Effective managers "age" their accounts receivable to ensure that they understand fully the amount of money owed to their businesses and to ensure that their receivables are, in fact, collected in a timely manner.
- Before extending (or continuing to extend) credit terms to a specific customer or guest, managers should consider:
  - The size of the customer's normal purchase and the potential for future business
  - The length of time the customer (if a business) has been in business
  - The status (age and balance) of the present account
  - The amount of time until the sale will be made
  - Where the customer falls on the operation's credit risk estimation
  - Whether a partial deposit should be required to extend credit
- Credit sales (if ultimately paid for by the guest!) are a positive occurrence for most businesses because these sales will eventually generate cash. **Write-offs**, which are in essence, official declarations that an account receivable is uncollectible, however, should be avoided whenever possible.

Cash and Accounts Payable (AP) Activities

- For many hospitality businesses, and especially for those who generate little or no credit sales (accounts receivable), the manipulation of accounts payable balances is the most commonly used method of managing temporary cash shortages. Simply put, when cash is in short supply and employees are at risk of not being able to cash their paychecks, the temporary postponement of normal bill paying may be necessary to ensure continued operations.
- Reliance on an untimely bill payment strategy to ease cash shortages is never a good long term strategy and an over-reliance on it is a significant sign that the business is either unprofitable or **undercapitalized,** which means it is chronically short of the capital (money) it needs to sustain its operation.

Cash Receipts/Disbursements Approach to Cash Budgeting

- When managers budget cash, they typically do so using the **cash receipts/disbursements approach to cash budgeting**, which requires managers to sum their cash receipts during a specific accounting period and then subtract

the cash amounts they will spend. The remaining balance represents their forecasted cash excess or cash shortage.
- See Figure 11.7 for an example of a typical cash budget. For a discussion of this cash budget, see *Go Figure!* in the text.

Managing Budgets through Internal Control

- Operations budgets help managers plan for revenues, expenses, and profits, while cash budgets help managers plan for cash availability. All the planning in the world, however, does not ensure that these budgets are implemented properly. Many internal and external forces change the way operations are carried out.
- In many cases, the safekeeping of revenue earned is most threatened *after* it is collected. This is because the risk of **theft,** or the unlawful taking of a business's property and **fraud**, the intentional use of dishonest methods to take property, is actually greatest from a business's own employees. In fact, for most hospitality businesses the risk of losing assets like cash from **embezzlement** (employee theft) is actually much greater than the risk of **robbery** (theft using force).
- It is very important for you to design and maintain control systems that ensure the security of your business's assets.
- Internal controls can be of a variety of types, but the internal control system itself seeks to:
    - Ensure accurate financial records keeping
    - Restrict unnecessary and potentially detrimental access to the business's assets
    - Confirm, periodically, that those responsible for safeguarding assets can account for them
    - Establish appropriate action steps for measuring and addressing variation between the expected and actual performance of those preserving the assets of the business
- An effective management control system for restaurants and hotels will have five fundamental characteristics. These consist of:
    - Management's concern for assets
    - Accurate data collection and comparison to written standards
    - Separation of responsibilities
    - Cost effectiveness
    - Regular external review

Management's Concern for Assets

- A fundamental characteristic of any effective control system is its emphasis on protecting company assets. This characteristic encompasses:
    - The documentation, collection and preservation of sales revenues
    - The care and safeguarding of assets such as equipment and inventories that are owned by the business
    - The careful documentation and payment of the business's legitimate expenses

Accurate Data Collection and Comparison to Written Standards

- An effective control system is one in which financial data is carefully recorded and then compared to previously identified standards or expectations.
- For example, assume that a business seeks to control part of its revenue collection process by monitoring the amount of money its cashiers have in their cash drawers (banks) at the end of their shifts. If the cashier is **over** (has more money than anticipated in the cashier's bank) or **short** (has less money than anticipated in the cashier's bank) the amount of the variation from expectations is carefully recorded.
- If management is to make this process an effective part of the overall control system, the data recorded by the cashiers must, first and foremost, be accurate
- Committing goals, standards, and expectations to writing is a critical part of an effective internal control system.
- In each instance where management seeks to control or monitor revenues and expenses, the record keeping systems in place must be consistently accurate. These management standards must be committed to written policies, both for the sake of consistency and so that employees fully understand the performance expectations by which they will be evaluated.

Separation of Responsibilities

- Effective accounting systems require talented individuals who are responsible for very specific and separate tasks. This means that only in the case of **collusion** (the secret cooperation of two or more employees) could the potential for fraud exist.
- With a clear separation of accounting duties, an internal system of checks and balances can be created that, when followed, can significantly reduce the chance of employee fraud and theft. In an effective internal control system, each involved employee is a member of a team.
- By separating component parts of the control process, even small restaurants and hotels can improve the quality of their control systems. Employee and management involvement and education is a key part of this process.

Cost Effectiveness

- Good control systems are also cost effective. The costs incurred to design, implement, and maintain a control system must be considered at the same time the overall worthiness of the control system is evaluated.
- For example, assume a new time-card system that eliminates employees punching in and out on paper time cards is available to hospitality operations for $50,000. The system uses retina scanning, and eliminates completely the possibility of **buddy punching**, the method by which an employee uses another's time card to punch that second employee in or out. To make a good decision about this purchase, a manager must determine:
    - The potential magnitude of the operation's "buddy punching" problem
    - The cost of using the retina scanner to address the problem

- See *Go Figure!* in the text for an illustration of this decision process.

Regular External Review

- External review, or audit, is an important characteristic of all quality internal control systems. An external review of a business's internal control systems may be undertaken by an outside audit group, by upper management, or even by a team assembled by the business's own on-site managers and staff. The goal in all cases is to review the accounting, reporting, and control systems and procedures that have been established, and to make recommendations for future changes and enhancements that can lead to continuous improvement.
- Depending upon the specific corporate form and the size of the company, methods of securing short or long term capital (money) to help meet budget requirements include:
  o The sale of fixed assets
  o The issuance of additional stock
  o Borrowing from a short-term source
  o Borrowing from a long-term source
- The amount of money an operation actually should keep readily on hand must be sufficient to operate the business, but an excessive amount of un-invested money creates a very small (if any) return on investment for its owners.

Key Terms & Concepts Review
Match the key terms with their correct definitions.

1. Budget _____
2. Chief Executive Officer (CEO) _____
3. Horizon _____
4. Long-range budget _____
5. Annual budget _____
6. Achievement budget _____
7. Operations budget _____
8. Cash budget _____

a. Having more money than anticipated in the cashier's bank.
b. A difference in dollars or percentage between budgeted and actual operating results that warrants further investigation.
c. An official declaration that an account receivable is uncollectible.
d. The difference between planned and actual results when actual results do not meet budget expectations (revenues are lower or expenses are higher).
e. The highest ranking officer in charge of the overall management of a company
f. The intentional use of dishonest methods to take property.
g. The secret cooperation of two or more employees to commit fraud
h. The unlawful taking of a business's property.

9. Capital expenditure _____ i. The method by which an employee uses another's time card to punch that second employee in or out.

10. Capital budget _____ j. The expense associated with the purchase of land, property and equipment, and other fixed assets that are recorded on the balance sheet.

11. Line item _____ k. A budgeting approach that sums the anticipated cash receipts and subtracts the anticipated cash payments during a specific accounting period to forecast cash excesses or cash shortages.

12. Payroll allocations _____ l. A report used by management to monitor the average length of time money owed to the business has remained uncollected.

13. Variance _____ m. A term used to describe a business that is chronically short of the capital (money) it needs to sustain its operation.

14. Favorable variance _____ n. Theft using force.

15. Unfavorable variance _____ o. A budget developed to estimate the actual impact on cash balances that will result from a business's operating activities, investing activities, and financing activities.

16. Significant variance _____ p. The concept of total cash receipts (cash in) and cash disbursements (cash out) which occurs in a business in a specific accounting period.

17. Flexible budget _____ q. Employee theft.

18. Cash budgeting _____ r. The difference between planned results and actual results.

19. Cash flow _____ s. Having less money than anticipated in the cashier's bank.

20. Accounts receivable aging report _____ t. A budget concerned with planning for the revenues, expenses, and profits associated with operating a business.

21. Write-offs _____ u. A budget used to plan for capital expenditures.

22. Undercapitalized _____ v. A budget prepared for a period of up to five years.

23. Cash receipts/ disbursements approach to cash budgeting _____ w. The difference between planned and actual results that are an improvement on the budget (revenues are higher or expenses are lower).

24. Theft _____ x. A budget prepared yearly.

25. Fraud _____   y. The general term used by managerial accountants to identify a variety of cash monitoring and management activities

26. Embezzlement _____   z. Financial plan.

27. Robbery _____   aa. Expense.

28. Over (cash bank) _____   bb. A budget that incorporates the assumptions of the original budget, such as fixed costs and target variable costs per unit or variable cost percentages, and then projects these costs based on varying levels of sales volume.

29. Short (cash bank) _____   cc. A short-range budget consisting of a month, a week, or a day.

30. Collusion _____   dd. The non-wage costs associated with, or allocated to, payroll.

31. Buddy punching _____   ee. Budget length into the future.

Discussion Questions

1. List and briefly describe three types of budgets, based on length.

2. List and briefly describe three types of budgets, based on purpose.

3. Explain the four steps in the operations budget monitoring process.

4. List the goals of an internal control system.

Quiz Yourself

Choose the letter of the best answer to the questions listed below.

Questions 1 – 10 are based on the information given below for Cheryl's Cookie Shack, owned by Cheryl Feeney, which is open in the summers on a beach in Rhode Island.

Budgeted Revenues for last year	$21,000.00
Budgeted Expenses for last year	$15,000.00
Sales last year	$22,500.00
Expenses last year	$14,685.50
Sales per cover last year	$4.50
Total cost per cover last year	$2.94
Food cost per cover last year	$1.30

1. What was the budgeted profit for last year for Cheryl's Cookie Shack?
   a. $6,000.00
   b. $4,300.00
   c. $7,814.50
   d. $7,500.00

2. Cheryl estimates that her sales will increase by 6.7% this year. What is her sales forecast?
   a. $22.407.00
   b. $16,005.00
   c. $19,725.30
   d. $24,007.50

3. Given a target food cost of 30%, what is Cheryl's forecasted food cost for this year?
   a. $6,722.10
   b. $7,202.25
   c. $4,801.50
   d. $6,750.00

4. If Cheryl estimates a food cost increase of 3%, what is her food cost per cover forecast for this year?
   a. $3.03
   b. $0.96
   c. $1.34
   d. $1.26

5. Given a standard labor cost of 32%, what is Cheryl's forecasted labor cost for this year?
   a. $5,121.60
   b. $7,170.24
   c. $6,312.10
   d. $7,682.40

6. Cheryl does not expect last year's payroll allocation of 9.5% to change this year. Based on her forecasted labor cost, what should she budget for payroll allocations?
   a. $486.55
   b. $729.83
   c. $681.17
   d. $599.65

7. What is Cheryl's total budgeted payroll for this year?
   a. $6,952.57
   b. $7,851.41
   c. $6,911.75
   d. $5,608.15

8. If Cheryl pays her brother a salary to manage her Cookie Shack this year, and the total for this, with payroll allocations, comes to $2,628.00, what is her budgeted hourly payroll?
   a. $5,054.40
   b. $5,223.41
   c. $4,324.57
   d. $4,283.75

9. What was Cheryl's expense variance based on her budget for last year?
   a. $6,000.00
   b. $1,500.00
   c. ($ 314.50)
   d. $6,314.50

10. What was Cheryl's percentage expense variance based on her budget for last year?
    a. (2.1%)
    b. 40.0%
    c. 10.2%
    d. 1.0%

Chapter 12

Capital Investment, Leasing, and Taxation

Highlights

* Capital Budgeting
* Capital Investment
* Time/Value of Money
* Rates of Return
* Capitalization Rates
* Financing Alternatives
* Debt vs. Equity Funding
* Lease vs. Buy Decisions
* Taxation
* Key Terms and Concepts Review
* Discussion Questions
* Quiz Yourself

Study Notes

Capital Budgeting

- In business, **capital** simply refers to money. Those who invest their capital are, not surprisingly, called **capitalists**, and the economic system that allows for the private ownership of property is called **capitalism.** As is the case in most industries, investing money in hospitality businesses can be risky.
- To achieve reasonable returns on their capital investment, owners must operate their businesses successfully; however, successful businesses are most often the end result of successful business planning.
- Operations budgets are concerned primarily with planning for the normal revenues and expenses associated with the day-to-day operating of a business. In the hospitality industry, capital budgets are used to plan and evaluate purchases of fixed assets such land, property, and equipment. Purchases of this type are called **capital expenditures** and, as you learned previously, are recorded on a business's balance sheet (recall that operations budget expenses are normally recorded on the income statement).
- **Capital budgeting** is simply the management process of evaluating the wisdom of one or more capital expenditures. These capital expenditures typically are more costly than those related to daily operating expenses, and thus, the owners or directors of a business pay particularly close attention to them.

- Capital budgeting is the essential process by which those in business evaluate which hospitality operations will be started, which will be expanded, and which will be closed.
- In nearly all cases, business owners seek returns on their investments which are large enough to justify the continued investment of their capital. In general, capital budgeting techniques can be classified as those that are directed toward one or more of the following business activities:
  o Establish a business
  o Expand a business (increase revenues)
  o Increase efficiency (reduce expenses)
  o Comply with the law
- **Establish a business.** Perhaps the most significant investment decision a person or a business entity can make is that of starting a new **venture** (new business). **Venture capitalists** are individuals or companies that are willing to take risks by financing promising new businesses.
- **Expand a business (increase revenues).** Expansion may take the form of building additional restaurants or hotels, increasing the capacity of an existing restaurant or hotel, or simply funding the extension of a single property's additional hours or services.
- **Increase efficiency (reduce expenses).** Capital budgeting decisions that result in updated facilities and/or equipment can significantly increase the productivity of a hospitality facility's workforce, thus reducing labor costs, and as a result, increasing profits
- **Comply with the law.** Mandated change may come from the passage of local, state, or federal regulations. When it does, a business must comply or face penalties and fines.
- Capital budgeting primarily addresses the funding of capital expenditures, while operations budgeting primarily addresses generating profits through operations.

Capital Investment

- Investors seek to balance the concepts of **risk**, (the likelihood that the investment will decline in value) with that of **reward** (the likelihood that the investment will increase in value). The two concepts are highly correlated. In most cases, as the amount of risk involved in an investment increases, the return on that investment also increases.
- To further illustrate the risk and reward relationship, assume, for example, that you had $1,000,000 to invest for one year. Assume also that you had identified four desirable investment options. The options are:
  1. Keep the money in a very safe place at your home.
  2. Deposit the money in a savings account at your local bank.
  3. Purchase $1,000,000 worth of stock in a publicly traded restaurant or hotel company.
  4. Invest the $1,000,000 to start your own restaurant or hotel.
- Figure 12.1 summarizes what has historically been the long-term risk/reward relationship found in the four investment options you have selected.

- For a discussion of the potential return on investment of the four options, see *Go Figure!* in the text.
- As an investor you would ultimately seek to compare the cost of making an investment today against the stream of income that the investment will generate in the future. To best make this "in the future" value comparison, it is important that you, and all investors, understand the **time value of money**, which is the concept that money has different values at difference points in time.

Time Value of Money

- To illustrate the time value of money concept, assume that you have won $10,000 in the state lottery. Your options for collecting payment are:
  - Receive $10,000 now, *or*
  - Receive $10,000 in four years.
- If you are like most people, you would choose to receive the $10,000 now. It makes little sense to **defer** (delay) a cash flow into the future when you could have the exact same amount of money now. At the most basic level, the time value of money simply demonstrates that, all things being equal, it is better to have money now rather than later.

Go Figure!

The total investment value formula is as follows:

Investment + Return on Investment = Total Investment Value

The value of the money that is invested now at a given rate of interest and grows over time is called the **future value** of money. The process of money earning interest and growing to a future value is called **compounding**.

If an investment is maintained for four years, it would grow as follows:

Year 1	$1,000 + ($1,000 X 0.10) = $1,100
Year 2	$1,100 + ($1,100 X 0.10) = $1,210
Year 3	$1,210 + ($1,210 X 0.10) = $1,331
Year 4	$1,331 + ($1,331 X 0.10) = $1,464

- The effect of compound investment returns is summarized in Figure 12.2.

Go Figure!

The formula managerial accountants use to quickly compute the future value of an investment when the rate of return and length of the investment is known is as follows:

> **Future Value = Investment Amount X (1 + Investment Earnings %)^n**
>
> *or*
>
> $FV_n = PV \times (1+i)^n$
>
> **Where FV equals the amount of the investment at the end of the investment period (future value), PV equals the present value of the investment, n equals the number of years the investment will be maintained, and i equals the interest rate %.**

In the example of a four-year investment, the future value formula would be computed as:

> $\$1,000 \times (1 + 0.10)^4$ = Future Value
>
> *or*
>
> $\$1,000 \times (1.464) = \$1,464$

- When a future value is known, then the **present value**, or the amount the future value of money is worth today, can be determined. The process of computing a present value is called **discounting**, or calculating the value of future money discounted to today's actual value. The formula used to quickly compute the present value of an investment when the rate of return and length of the investment is known is as follows:

> **Present Value = $\dfrac{\text{Future Value}}{(1 + \text{Investment Earnings \%})^n}$**
>
> *or*
>
> $PV = \dfrac{FV_n}{(1+i)^n}$
>
> **Where FV equals the amount of the investment at the end of the investment period (future value), PV equals the present value of the investment, n equals the number of years the investment will be maintained, and i equals the interest rate %.**

In the example of a four-year investment, the present value formula would be computed as:

> $\dfrac{\$1,464}{(1+0.10)^4}$ = Present Value
>
> *or*
>
> $\dfrac{\$1,464}{1.464} = \$1,000$

183

Put another way, the $1,464 investment (received four years from now) would be worth $1,000 today.

- Future values and present values can be calculated using the formulas stated in this chapter, time value of money tables, and/or financial calculators.
- As you (and all savvy investors) now recognize, maximum returns on money invested (ROI) are achieved by utilizing one or both of the following investment strategies:
  1. Increasing the length of time money is invested
  2. Increasing the annual rate of return on the investment
- The effect of these two variables on investment returns is shown graphically in Figure 12.3.

Rates of Return

- Before closely examining rates of return, it is very important for those in the hospitality industry (as well as all other industries!) to understand that operating profits are not the same as return on investment.
- Sometimes, a restaurant that achieves a very good profit (net income) is still not a good investment for the restaurant's owner. In other cases, a restaurant that achieves a less spectacular net income is a better investment.

Go Figure!

Using the ROI formula you learned about in Chapter 3, the owners' ROIs can be calculated as follows:

$$\frac{\text{Money Earned on Funds Invested}}{\text{Funds Invested}} = \text{ROI}$$

- Actual returns on investment can vary greatly, but few, if any, investors will for a long period of time invest in a restaurant if the net income is less than what could be achieved in other investment opportunities with the same or lesser risks.
- Sophisticated managerial accountants can utilize several variations of this basic ROI formula to help them make good decisions about investing their capital. For working managers interested in maximizing returns on investment, two of the most important of these formula variations are:
  o Savings Rate of Return
  o Payback Period

Savings Rate of Return

- The **savings rate of return** is the relationship between the annual savings achieved by an investment and the initial capital invested. To compute the estimated savings rate of return on a proposed capital expenditure for a dish machine, for example, you must first collect some important information, which includes:
  - The **book value** of the existing dish machine (the machine value as listed on the balance sheet)
  - The life expectancy of the current machine
  - The value (if sold) of the existing machine
  - The annual operating costs of the current machine
- Similar information must be obtained for the new piece of equipment. Thus, you must determine:
  - The purchase price (including installation) of the new machine
  - The life expectancy of the new machine
  - The value (when ultimately sold) of the new machine
  - The annual operating costs of the new machine
- An example of this information is presented in Figure 12.4.

Go Figure!

Based on the data from Figure 12.4, you can compute your savings rate of return as follows:

$$\frac{\text{Annual Savings}}{\text{Capital Investment}} = \text{Savings Rate of Return}$$

- In many cases, managerial accountants and/or the owners of a business will set an investment **return threshold** (minimum rate of return) that must be achieved prior to the approval of a capital expenditure.

Payback Period

- **Payback period** refers to the length of time it will take to recover 100% of an amount invested. Typically, the shorter the time period required to recover all of the investment amount, the more desirable it is.

Go Figure!

In the dish machine example cited in Figure 12.4, the payback period is computed as:

$$\frac{\text{Capital Investment}}{\text{Annual Income (or Savings)}} = \text{Payback Period}$$

- Managerial accountants often utilize the payback period formula to evaluate different investment alternatives.

Capitalization Rates

- You have learned how business investors calculate their ROI when the amount earned on an investment is known. In most cases, however, business investors are not guaranteed a return on their investments. Rather, these investors must estimate the returns they will achieve prior to making their investment decisions.
- Investment returns typically increase as an investment's risk level increases. The concept of risk and aversion to (or avoidance of) a specific risk level is a very personal one. Some businesses succeed when few believed they would, while other businesses seem poised for success and then fail. In the final analysis, all lenders or investors must be comfortable with the risk level of their investments and will want to see a realistic estimate of their return on investment (ROI) prior to investing.
- To fully appreciate how business investors estimate their ROIs and then consider risk levels, you must first understand capitalization rates. In the hospitality industry, **capitalization (cap) rates** are utilized to compare the price of entering a business (the investment) with the anticipated, but not guaranteed, returns from that investment (net operating income). The computation for a cap rate is:

$$\frac{\text{Net Operating Income}}{\text{Investment Amount}} = \text{Cap Rate \%}$$

- This formula directly ties investment returns to:
  - The size of the profits (net operating income) generated by the business
  - The size of the investment in the business
- **Net operating income (NOI)**, in general, is the income before interest and taxes you would find on a restaurant or hotel income statement (see Chapter 3).
- Investors generally do not want to pay more than the true value of any specific hospitality business or property they are considering purchasing.

Go Figure!

The cap rate % formula can be restated using property value as the investment amount. The computation of the cap rate in this case would be:

$$\frac{\text{Net Operating Income}}{\text{Property Value}} = \text{Cap Rate \%}$$

- Cap rates that are *higher* tend to indicate a business is creating very favorable net operating incomes relative to the business's value (selling price). Cap rates that

are *lower* indicate that the business is generating a smaller level of net operating income relative to the business's estimated value (selling price).
- In general, cap rates are used to indicate the rate of return investors expect to achieve on a known level of investment. By using the rules of algebra, investors can modify the formula to help them determine the value, or purchase price, of a specific hospitality business.

Go Figure!

The property value estimate is calculated as follows:

$$\frac{\text{Net Operating Income}}{\text{Cap Rate \%}} = \text{Property Value Estimate}$$

- There are two areas of common confusion when computing cap rates. The first difficult area is the definition of net operating income itself, and the second is the accounting period on which the net operating income is based.
- The decision to include and how to include the actual expenses of the business will, of course negatively or positively affect the overall net operating income achieved and, as a result, will directly affect the calculated cap rate.
- In addition to net operating income, the accounting period analyzed will also significantly impact the cap rate computation. Net operating income for a single (very good or very bad) year may or may not be a true reflection of a business's actual ability to create a consistent flow of operating income. Because this is true, it is normally best to analyze several years of net operating income when that is possible, with a particular concentration on whether net operating income is increasing, staying the same, or declining each year.
- The importance of doing so can be demonstrated by the advertisement created by the owners of a hotel who now wish to **flip** (sell) the property (see Figure 12.5).
- See *Go Figure!* in the text for a demonstration of the difference in the cap rate % when it is computed using an average rate for several years, and the result when only one year's performance is used.
- Consider the case of a hotel which has been poorly managed and therefore has a poor net operating income level. If the property were up for sale, the seller would likely take the position that the hotel has great **upside potential** (possible increased future value). A potential investor (buyer), however, would likely seek to purchase the hotel at the lowest possible price, and thus would take the position that improved management may indeed improve the hotel, but no guarantee exists that it would do so.
- Traditionally, hospitality business values have been established through the use of one or more of the following evaluation methods.
- **Replacement approach.** This approach assumes that a buyer would not be willing to pay more for a business than the amount required to build (replace) it with a similar business in a similar location.

- **Revenue Approach.** This approach views a hospitality business primarily as a producer of revenue. Thus, a business's value is established as a multiple of its annual revenue.
- **Capitalization Approach.** This system seeks to develop a mathematical relationship (cap rate) between a business's projected net operating income and its market value. This approach, while often the most complex to employ, is also the most widely used.

Financing Alternatives

- For investors, **financing** simply refers to the method of securing (funding) the money needed to invest.
- Theoretically, financing alternatives for a purchase could range from paying cash for the full purchase price (100% equity financing) to borrowing the full purchase price (100% debt financing).
- Because investments are typically financed with debt and/or equity funds, the precise manner in which financing is secured will have a major impact on the return on investment (ROI) investors ultimately achieve.

Debt vs. Equity Financing

- With **debt financing**, the investor borrows money and must pay it back with interest within a certain timeframe. With **equity financing**, investors raise money by selling a portion of ownership in the company. If investors use their own money for equity financing, they are, in effect, selling the ownership in the business to themselves.
- Common suppliers of debt financing include banks, finance companies, credit unions, credit card companies, and private corporations. An advantage of debt financing is that, as investors pay down their loans, they also build creditworthiness. This makes them even more attractive to lenders and increases their chances of negotiating favorable loan terms in the future.
- Equity financing typically means taking on investors and being accountable to them. Many small business investors secure equity funds from relatives, friends, colleagues, or customers who hope to see their businesses succeed and get a return on their investment. Other sources of equity financing can include venture capitalists, individuals with substantial net worth, corporations, and financial institutions. **Passive investors** are willing to give capital but will play little or no part in running the company, while **active investors** expect to be heavily involved in the company's operations.
- Regardless of whether an investor funds an investment with their own equity, with equity from passive investors, or with equity from active investors, a ROI on equity funds is achieved only *after* those who have supplied debt funding have earned their own ROIs. Despite this fact, equity financing is not free. Equity investors typically are entitled to a share of the business's profits as long as they hold, or maintain, their investments.

- The amount of ROI generated by an investment is greatly affected by the ratio of debt to equity financing in that investment. The debt to equity ratio in an investment will also affect the willingness of lenders to supply investment capital.

Go Figure!

If an owner invested all of his/her own money, her ROI would be computed as follows:

$$\frac{\text{Money Earned on Funds Invested}}{\text{Funds Invested}} = \text{ROI}$$

- Figure 12.6 illustrates the affect on equity ROI of funding the investment with varying levels of debt and equity.
- For any investment, the greater the financial leverage (see Chapter 6) or funding supplied by debt, the greater the ROI achieved by the investor. As you can see from the data in Figure 12.6, if the owner elects to fund 50% of the project's cost with equity and 50% with debt, the ROI is much higher than if he/she did not leverage her investment at all. With even greater leverage (for example, 80% debt financing and 20% equity financing), there are even greater investment returns on the equity investment.
- Given the data presented in Figure 12.6, investors may ask, "Why not fund nearly 100% of every investment using debt?" The answer to this question lies in the column titled Debt Coverage Ratio. This is the same ratio as the Times Interest Earned ratio that you learned about in Chapter 6.

Go Figure!

The debt coverage ratio can be calculated as follows:

$$\frac{\text{Net Operating Income}}{\text{Interest Payment}} = \text{Debt Coverage Ratio}$$

- The **debt coverage ratio** is a measure of how likely the business is to actually have the funds necessary for loan repayment. Just as investors will maintain minimum ROI thresholds prior to considering an investment to be a "good" one, lenders will analyze debt coverage ratios and their own thresholds to determine the risk they are willing to assume when lending to an investor's project.
- Those who seek debt funding, but whose projects show lower than desirable debt coverage ratios will most often find that the only money lent to their project will come at a higher cost.
- Because many lending institutions consider businesses in the hospitality industry to be high risk, some will provide debt financing for no more than 50 to 70% of a project's total cost. This creates a **loan to value (LTV) ratio**, a ratio of the outstanding debt on a property to the market value of that property, of 50- 70%,

while expecting the project's owner/investor(s) to secure the balance of the project's cost in equity funding.

Lease vs. Buy Decisions

- Leasing, (an agreement to **lease**) allows a business to control and use land, buildings, or equipment without buying them. Just as a business's decision to vary the debt/equity ratio funding of a project will directly affect its investors' ROI, so too will ROI be affected by a business entity's decision to lease rather than purchase capital assets.
- Leasing can provide distinct advantages for the **lessor** (the entity which owns the leased property) as well as for the **lessee** (the entity that leases the asset). In a lease arrangement, lessors gain immediate income while still maintaining ownership of their property. Lessees enjoy limited property rights and distinct financial and tax advantages, as well as the right, in many cases, to buy the property at an agreed upon price at the end of the lease's term.
- From a legal perspective, as well as a financial perspective, buying an asset is much different than leasing it. Figure 12.7 details some significant differences between the rights of property owners (lessors) and lessees of that same property.
- The most significant financial difference between buying and leasing is that payments for *owned* property are not listed as a business expense on the monthly income statement. Rather, the value of the asset is depreciated over a period of time appropriate for that specific asset. Payments for most (but not all) *leased* property, however, are considered an operating expense and thus are listed on the income statement.
- Leasing capital equipment has become increasingly popular for a variety of reasons including advantages listed in Figure 12.8.
- Despite their varied advantages, leases can have distinct disadvantages in some cases. These may include:
  - Non-ownership of the leased item or property at the end of the lease.
  - In situations where few lease options exist, the cost of leasing may ultimately be higher than the cost of buying.
  - Changing technology may make leased equipment obsolete but the lease term is unexpired.
  - Significant penalties may be incurred if the lessee seeks to terminate the lease before its original expiration date.
- Perhaps the most significant decision to be made prior to obtaining a lease on a capital asset such as equipment is simply that of ultimate ownership. That is, does the organization seeking the lease want to ultimately own the leased item? If so, a purchase of the asset may be a better option.

Taxation

- Managerial accountants who fully understand the very complex tax laws under which their businesses operate can help ensure that the taxes these businesses pay are exactly the amount owed by the business, and no more. This is important because when the correct amount of tax is paid, ROIs will be maximized and not wrongfully reduced.

Income Taxes

- Taxing entities such as the federal, state, and local governments generally assess taxes to businesses based upon their own definitions of taxable income. It is important to understand that income taxes are not voluntary. However, no business should be required to pay *more* taxes than it actually owes. It is the job of the tax accountant to ensure that the businesses they advise do not overpay.
- It is important to recognize the difference between tax avoidance and tax evasion. **Tax avoidance** is simply planning business transactions in such a way as to minimize or eliminate taxes owed. **Tax evasion**, on the other hand, is the act of reporting inaccurate financial information or concealing financial information in order to evade taxes by illegal means. Tax avoidance is legal and ethical, while tax evasion is not.
- **Taxable income** is generally defined as gross income adjusted for various deductions allowable by law. There are regulations that must be followed in computing taxes, and a company's taxable income will vary based upon a variety of factors, including a specific taxing authority's unique definition of taxable income.
- One of the first questions asked by a **tax accountant**, a professional who assists businesses in paying their taxes using standard accounting practices and current tax laws, is that of organizational structure. This is so because different business structures are taxed differently.
- The most popular form of business entity is the S corporation. Those who choose the S corporation typically make the decision to do so because it provides limited liability, fairly simple tax filings, and possible tax advantages at the state level. Another popular form of business entity is the partnership.
- There are also about 2.1 million regular, or "C" corporations in the United States. The income tax rates (at the time of this book's publication) for C corporations are shown in Figure 12.8.
- When most hospitality business owners and their tax accountants examine their taxes, the first tax to be considered is the income tax. It is important to understand that all business's net income is taxed at the federal level, but it most often also is taxed at the state and even local levels.

Capital Gains Taxes

- A **capital gain** is the surplus that results from the sale of an asset over its original purchase price adjusted for depreciation (**asset basis**). A **capital loss** occurs when the price of the asset sold is less than the original purchase price adjusted for depreciation.
- Capital gains and losses occur with the sale of real assets, such as property, as well as financial assets, such as stocks and bonds. The federal government (and some states) imposes a tax on gains from the sale of assets. **Realized capital gains** occur when the actual sale of the asset is completed. The difference between capital gain and capital loss is illustrated in *Go Figure!* in the text.

Property Taxes

- In addition to income and capital gains taxes, most hospitality businesses will be responsible for paying property taxes on the property owned by their businesses. These assessments are determined through a **real estate appraisal**, which is an opinion of the value of a property, usually its market value, performed by a licensed appraiser. **Market Value** is the price at which an asset would trade in a competitive setting. For a hypothetical property tax bill, see Figure 12.9.

Other Hospitality Industry Taxes

- In addition to income taxes, capital gains taxes, and property taxes that apply to all businesses, hospitality managers and their tax accountants are responsible for reporting and paying a variety of hospitality industry taxes. Three of the most important of these taxes include:
- **Sales Tax.** In most cases, **sales taxes** are collected from guests by hospitality businesses for taxes assessed on the sale of food, beverages, rooms, and other hospitality services. Typically, the funds collected are then transferred (forwarded) to the appropriate taxing authority on a monthly or quarterly basis.
- **Occupancy (Bed) Tax.** Occupancy (bed) taxes are a special assessment collected from guests and paid to a local taxing authority based upon the amount of revenue a hotel achieves when selling its guest rooms.
- **Tipped Employee Tax. Tipped employee taxes** are assessed on tips and gratuities given to employees by guests or the business as taxable income for those employees. As such, this income must be reported to the IRS, and taxes, if due, must be paid on that income.

Modified Accelerated Cost Recovery System (MACRS)

- Depreciation is a method of allocating the cost of a fixed asset over the useful life of the asset. Depreciation is subtracted on the income statement primarily to lower income, thus lower taxes. The portion of assets depreciated each year is

considered "tax deductible" because it is subtracted on the income statement before taxes are calculated.
- The **Modified Accelerated Cost Recovery System (MACRS)** is the depreciation method required for equipment in the hospitality industry (and all industries). MACRS was designed to accelerate depreciation in the first years of depreciating an asset in order to reduce the amount of taxes paid in those years. Depreciation for real estate, however, follows **straight-line depreciation**, which is the cost of the asset divided evenly over the life of the asset.
- MACRS establishes shorter recovery periods than in straight-line depreciation. Also, calculations are based on property class lives and an estimated **salvage value** (the estimated value of an asset at the end of its useful life) of zero. Property class lives as they are applied to hospitality businesses are shown in Figure 12.10.
- Depreciation using MACRS is calculated using stated percentages for varying property classes (see Figure 12.11). You may notice that the number of **recovery years** (years of depreciation for the asset) include one more year than the years of the property class. This is because of the half-year convention. The **half-year convention** allows for one-half of a year's depreciation to be taken in the year of purchase and one-half in the year following the end of the class life. In effect, this allows for one more year of depreciation to occur.
- If an asset has a known salvage value, then the salvage value would be subtracted from the original cost of the asset before depreciation is calculated. See *Go Figure!* in the text for an illustration of the application of MACRS.
- While it is not reasonable for most hospitality managers to become tax experts, it is possible for them to:
  1. Be aware of the major entities responsible for tax collection and enforcement.
  2. Be aware of the specific tax deadlines for which they are responsible.
  3. Stay abreast, to the greatest degree possible, of changes in tax laws that may directly affect their business.
- The Internal Revenue Service (IRS) is the taxing authority with which hospitality managers are likely most familiar.
- Among other things, the IRS requires businesses to:
  - File quarterly income tax returns and make payments on the profits earned from business operations.
  - File an Income and Tax Statement with the Social Security Administration.
  - Withhold income taxes from the wages of all employees and deposit these with the IRS at regular intervals.
  - Report all employee income earned as tips and withhold taxes on the tipped income.
  - Record the value of meals charged to employees when the meals are considered a portion of an employee's income.
  - Furnish a record of withheld taxes to all employees on or before January 31 of each year (Form W-2).
- It is the role of hospitality managers to stay abreast of significant changes in tax laws and follow them to the letter.

Key Terms & Concepts Review

Match the key terms with their correct definitions.

1. Capital _____
2. Capitalist _____
3. Capitalism _____
4. Capital expenditures _____
5. Capital budgeting _____
6. Venture _____
7. Venture capitalist _____
8. Risk _____
9. Reward _____
10. Time value of money _____
11. Defer _____
12. Future value _____
13. Compounding _____
14. Present value _____
15. Discounting _____
16. Savings rate of return _____
17. Book value _____

a. Gross income adjusted for various deductions allowable by law.
b. The money that a business must collect from customers and pay to taxing authorities as a result of realizing taxable sales.
c. The process of money earning interest and growing to a future value.
d. The surplus that results from the sale of an asset over its original purchase price adjusted for depreciation when the actual sale of an asset is completed.
e. A depreciation method in which the cost of the asset is divided evenly over the life of the asset.
f. Funding an investment by borrowing money and paying it back with interest within a certain timeframe.
g. The value of money that is invested now at a given rate of interest and grows over time.
h. The entity that leases an asset.
i. Individual or company that is willing to take risks by financing promising new businesses.
j. The loss that results when the price of an asset sold is less than the original purchase price adjusted for depreciation.
k. An opinion of the value of a property, usually its market value, performed by a licensed appraiser.
l. A term meaning to sell.
m. Taxes assessed on tips and gratuities given to employees by guests or the business as taxable income for those employees.
n. An economic system that allows for the private ownership of property.
o. The surplus that results from the sale of an asset over its original purchase price adjusted for depreciation.
p. A MACRS depreciation method principle that allows for one-half of a year's depreciation to be taken in the year of purchase and one-half in the year following the end of the class life.
q. The purchase price of an asset minus accumulated depreciation.

18. Return threshold _____
19. Payback period _____
20. Capitalization (cap) rates _____
21. Net operating income (NOI) _____
22. Flip _____
23. Upside potential _____
24. Financing _____
25. Debt financing _____
26. Equity financing _____
27. Passive investors _____
28. Active investors _____
29. Debt coverage ratio _____
30. Loan to value (LTV) ratio _____
31. Lease _____
32. Lessor _____
33. Lessee _____
34. Tax avoidance _____
35. Tax evasion _____
36. Taxable income _____
37. Tax accountant _____

r. The price at which an asset would trade in a competitive setting.
s. The entity which owns a leased property.
t. A ratio of the outstanding debt on a property to the market value of that property.
u. A term used for money.
v. Investors who expect to be heavily involved in the company's operations.
w. Years of depreciation for an asset when using the MACRS depreciation method
x. An agreement that allows a business to control and use land, buildings, or equipment without buying them.
y. The depreciation method required for equipment which accelerates depreciation in the first years of depreciating an asset in order to reduce the amount of taxes paid in those years.
z. The process of computing a present value or calculating the value of future money discounted to today's actual value.
aa. The expenses associated with the purchase of land, property and equipment, and other fixed assets that are recorded on the balance sheet.
bb. The estimated value of an asset at the end of its useful life.
cc. The income before interest and taxes found on a restaurant or hotel income statement.
dd. A professional who assists businesses in paying their taxes using standard accounting practices and current tax laws.
ee. The method of securing (funding) the money needed to invest.
ff. The amount a future value of money is worth today.
gg. The likelihood that an investment will increase in value.
hh. The value of the existing piece of equipment as it is listed on the balance sheet.
ii. A new business.
jj. Funding an investment by selling a portion of ownership in the company.
kk. The act of planning business transactions in such a way as to minimize or eliminate taxes owed.

38. Capital gain _____ ll. The concept that money has different values at difference points in time.

39. Capital loss _____ mm. The likelihood that an investment will decline in value.

40. Realized capital gains _____ nn. A person who invests money..

41. Asset's basis _____ oo. The management process of evaluating one or more capital expenditures

42. Real estate appraisal _____ pp. The possible increased future value of an investment.

43. Market value _____ qq. The minimum rate of return that must be achieved on an investment.

44. Sales taxes _____ rr. The length of time it will take to recover 100% of an amount invested.

45. Tipped employee taxes _____ ss. The act of reporting inaccurate financial information or concealing financial information in order to avoid taxes by illegal means

46. Modified Accelerated Cost Recovery System (MACRS) _____ tt. The rates utilized to compare the price of entering a business (the investment) with the anticipated, but not guaranteed, returns from that investment (net operating income).

47. Straight-line depreciation _____ uu. The relationship between the annual savings achieved by an investment and the initial capital invested.

48. Salvage value _____ vv. A measure of how likely a business is to have the funds necessary for loan repayment.

49. Recovery years _____ ww. Investors who are willing to give capital but will play little or no part in running the company.

50. Half-year convention _____ xx. To delay

Discussion Questions

1. List four business activities that are addressed by capital budgeting techniques.

   _____
   _____
   _____

2. List and briefly explain two important variations of the ROI formula used by managerial accountants.

3. Explain the difference between debt and equity financing.

4. List six types of taxes typically paid by hospitality businesses.

Quiz Yourself

1. Jerry has $25,000 to invest and he expects to receive an 8% annual rate of return. What will his total investment value be at the end of the first year?
    a. $ 28,125
    b. $ 35,000
    c. $ 27,000
    d. $ 26,250

2. What is the future value of Jerry's investment at the end of 4 years?
    a. $ 33,000.10
    b. $ 34,012.22
    c. $ 27,000.00
    d. $108,000.00

3. If Jerry wanted to have $40,000 at the end of the 4 years, and given an 8% rate of return, how much would he have to invest today?
    a. $23,145.15
    b. $33,000.10
    c. $25,925.93
    d. $29,401.19

4. Jerry's brother Jim invested $20,000 five years ago at 10% and has earned $12,210.20 on his investment. What is his ROI?
    a. 61.05%
    b. 204.75%
    c. 10.01%
    d. 40.00%

5. Jim bought a new shuttle van for his hotel for $35,000. It has an expected life of 7 years. If the new van represents an annual savings of $5,000, what is his savings rate of return?
   a. 7.00%
   b. 24.50%
   c. 700.00%
   d. 14.29%

6. What is the payback period for the new van?
   a. 60 months
   b. 20 years
   c. 7 years
   d. 5 years

7. Jerry is considering investing $750,000 in Jim's hotel, which has gross annual revenues of $1,200,000 and a net operating income of $90,000. Calculate the cap rate for his investment.
   a. 11.20%
   b. 12.00%
   c. 54.51%
   d. 43.50%

8. Jerry is also looking at another property in the area, which is valued at $850,000, and has a net operating income of $48,000. What will be his cap rate in this case?
   a. 5.65%
   b. 3.40%
   c. 7.65%
   d. 19.20%

9. Jerry hopes to achieve a 12.5% cap rate. At a third hotel, he finds the net operating income is $102,000. What should he offer for the property to meet his goals (that is, what would the property value be?)
   a. $1,275,000
   b. $3,672,000
   c. $ 816,000
   d. $ 750,000

10. Jim now needs to borrow money from the bank for his hotel. His net operating income is $90,000 and he will be paying 8.0% on a loan of $600,000 with an annual payment of $48,000. Calculate his debt coverage ratio.
    a. 0.816
    b. 2.500
    c. 0.630
    d. 1.875

Answers to Key Terms & Concepts Review, Discussion Questions, and Quiz Yourself

Chapter 1

Key Terms & Concepts Review

1. l	7. p	13. v	19. t	25. o	31. r
3. a	9. hh	15. b	21. i	27. ee	33. s
5. cc	11. q	17. f	23. n	29. gg	

Discussion Questions

1. Describe at least one way in which an owner, an investor, and a manager would use accounting in the hospitality industry.
   a. Owners want to monitor their business's financial condition
   b. Investors want choose a businesses that will conserve or increase their wealth
   c. Managers want to make the best management decisions possible for themselves and their businesses

3. Define a uniform system of accounts and list three systems which are used by the hospitality industry.
   - A uniform system of accounts represents agreed upon methods of recording financial transactions within a specific industry segment
   - USAR – used by the restaurant industry
   - USALI – used by the lodging industry
   - USFRC – used by the club industry

Quiz Yourself

1. b	3. d	5. d	7. b	9. c

Chapter 2

Key Terms & Concepts Review

1. v	11. z	21. xx	31. t	41. tt	51. uu
3. ss	13. bbb	23. nn	33. u	43. b	53. ll
5. p	15. q	25. o	35. e	45. j	55. w
7. c	17. s	27. hh	37. ddd	47. ee	
9. dd	19. l	29. bb	39. ccc	49. pp	

Discussion Questions

1. List five concepts to remember about maintaining a business's general ledger.
   - The Accounting Formula, which is the summary of a business's asset, liability and owners' equity accounts, must stay in balance.
   - The Accounting Formula is affected every time a business makes a financial transaction.
   - Each financial transaction is to be recorded two times in a double entry accounting system.
   - The original records of a business's financial transactions are maintained in its journal, and each financial transaction recorded is called a journal entry.
   - The current balances of each of a business's individual asset, liability and owners' equity accounts are totaled and maintained in its general ledger.

3. Describe the hand "trick" and list the accounts represented by each finger.
   - Your left hand represents debits and your right hand represents credits.
   - Your fingers represent the following accounts, where the left hand shows debits and the right hand shows credits:

Thumb -	Assets (A)
Pointer -	Liabilities (L)
Middle finger -	Owners' Equity (OE)
Ring finger -	Revenues (R)
Pinky finger -	Expenses (E)

   - Fingers "up" represent increases and fingers "down" represent decreases.

Quiz Yourself

1. d	3. d	5. c	7. b	9. c

Chapter 3

Key Terms & Concepts Review

1. i	9. dd	17. s	25. hh	33. mm	41. d
3. ll	11. j	19. qq	27. c	35. cc	43. r
5. t	13. z	21. jj	29. nn	37. h	45. k
7. n	15. b	23. p	31. o	39. ff	

Discussion Questions

1. List the five stakeholders affected by a business's profitability.
   - Owners
   - Investors
   - Lenders
   - Creditors
   - Managers

3. List the three sections of a USALI income statement.
   - Operated department income
   - Undistributed operating expenses
   - Nonoperating expenses

Quiz Yourself

1. a	3. b	5. a	7. b	9. c

Chapter 4

Key Terms & Concepts Review

1. i	7. l	13. ff	19. a	25. x	31. gg
3. q	9. cc	15. z	21. m	27. d	33. e
5. r	11. b	17. aa	23. bb	29. s	

Discussion Questions

1. List the five groups affected by the information contained in the balance sheet and briefly discuss what they are looking for on the balance sheet.
   - Owners – lets them know about the amount of the business which they actually "own".
   - Investors – they can measure the return on their investment.
   - Lenders – can better understand the financial strength (and thus the repayment ability) of the business
   - Creditors – like lenders, they are concerned about repayment
   - Managers – they must be able to read and analyze their own balance sheets to determine items such as the current financial balances of cash, accounts receivable, inventories, and accounts payable, and other accounts that have a direct impact on operations

3. List and briefly explain the three major components of the balance sheet.
   - Assets – all items owned by the business
   - Liabilities – all monies owed by the business to others
   - Owner's equity – the residual claims the owner's have on their assets

Quiz Yourself

1. b   3. a   5. b   7. d   9. a

Chapter 5

Key Terms & Concepts Review

1. d   7. f   13. q
3. n   9. i   15. j
5. o   11. a   17. c

Discussion Questions

1. Explain the reason cash flows are so critically important to the operation of a successful business.
   - Having access to cash at the right time (versus having the right amount of cash) is critically important to businesses.

3. List the five major components of the Statement of Cash Flows.
   - Cash flow from operating activities
   - Cash flow from investing activities
   - Cash flow from financing activities

- Net changes in cash
- Supplementary schedules

Quiz Yourself

1. c   3. a   5. d   7. b   9. c

Chapter 6

Key Terms & Concepts Review

1. mm	9. k	17. w	25. a	33. g	41. s
3. ee	11. b	19. y	27. p	35. vv	43. gg
5. ii	13. l	21. jj	29. c	37. e	45. h
7. x	15. aa	23. pp	31. rr	39. q	47. i

Discussion Questions

1. Use three methods of expressing percentages to express the value of 10 percent.
   - Fraction – 10/100
   - Decimal – 0.10
   - Percent – 10%

3. List five industry-specific ratios used by hotels.
   - Occupancy Percentage
   - Average Daily Rate (ADR)
   - Revenue Per Available Room (RevPAR)
   - Revenue Per Available Customer (RevPAC)
   - Cost Per Occupied Room (CPOR)

Quiz Yourself

1. d   3. a   5. c   7. b   9. b

Chapter 7

Key Terms & Concepts Review

1. j 3. h 5. l 7. c 9. b 11. d

Discussion Questions

1. List the factors that influence a foodservice operation's menu price.
   - Local Competition
   - Service Levels
   - Guest Type
   - Product Quality
   - Portion Size
   - Ambience
   - Meal Period
   - Location
   - Sales Mix

3. Describe the matrix analysis method, using food cost percentage.
   - To analyze a menu using the food cost percentage method, menu items must be segregated based on the following two variables:
     - Popularity (number sold)
     - Food cost percentage
   - You are seeking menu items that have the effect of *minimizing your overall food cost percentage*, since a lowered food cost percentage leaves more of the sales dollar to be spent for other operational expenses.

Quiz Yourself

1. a 3. b 5. d 7. b 9. c

Chapter 8

Key Terms & Concepts Review

1. w	9. ee	17. bb	25. ff	33. q	41. u
3. n	11. cc	19. y	27. j	35. z	
5. mm	13. kk	21. ii	29. nn	37. dd	
7. aa	15. hh	23. d	31. c	39. h	

204

Discussion Questions

1. List the seven steps required to complete the Hubbart room rate formula.
   - Calculate hotel's target before-tax net income
   - Calculate estimated non operating expenses
   - Calculate estimated undistributed operating expenses
   - Calculate estimated operated departments income, excluding rooms
   - Calculate the operated department income for rooms
   - Calculate the estimated rooms department revenues based on estimated occupancy
   - Calculate the hotel's require ADR

3. Define the concept of revenue management for a hotel.
   - Revenue management, also called yield management, is a set of techniques and procedures that use hotel specific data to manipulate occupancy, ADR, or both for the purpose of maximizing the revenue yield achieved by a hotel.

Quiz Yourself

1. a   3. a   5. d   7. a   9. d

Chapter 9

Key Terms & Concepts Review

1. l   5. i   9. c   13. d   17. f   21. g
3. p   7. n   11. u   15. v   19. o   23. m

Discussion Questions

1. Identify ten ways business costs can be classified.
   - Fixed and variable costs
   - Mixed costs
   - Step costs
   - Direct and indirect (overhead) costs
   - Controllable and non-controllable costs
   - Other costs:
     - Joint costs
     - Incremental costs
     - Standard costs
     - Sunk costs
     - Opportunity costs

3. List the three steps in a high/low analysis of costs.
- Determine variable cost per guest for the mixed cost.
- Determine total variable costs for the mixed cost.
- Determine the fixed costs portion of the mixed cost.

Quiz Yourself

1. c   3. b   5. d   7. c   9. b

Chapter 10

Key Terms & Concepts Review

1. i   5. h   9. n   13. b   17. a

3. k   7. l   11. m   15. c

Discussion Questions

1. Identify three basic truths about forecasts.
   - Forecasts involve the future
   - Forecasts rely on historical data
   - Forecasts are best utilized as a "guide"

3. List the reasons local property managers can make the best forecasts for their hotels.
   - They understand the unique property features affecting demand for their hotels.
   - They know about special city-wide events in the area that affect demand.
   - They understand the demand for competitive hotel properties in the area.
   - They can factor in the opening or closing of competitive hotels in the area.
   - They can include factors such as weather, road construction, and seasonality in their demand assessments.
   - They can adjust forecasts very quickly when faced with significant demand-affecting events (i.e., power outages and airport or highway closings).

Quiz Yourself

1. c   3. d   5. a   7. c   9. a

Chapter 11

Key Terms & Concepts Review

1. z	7. t	13. r	19. p	25. f	31. i
3. ee	9. j	15. d	21. c	27. n	
5. x	11. aa	17. bb	23. k	29. s	

Discussion Questions

1. List and briefly describe three types of budgets, based on length.
   - A long-range budget is typically prepared for a period of up to five years.
   - An annual budget, or yearly budget, is for a one-year period or, in some cases, one season.
   - An achievement budget, or short-range budget, is always of a limited time period, often consisting of a month, a week, or even a day.

3. Explain the four steps in the operations budget monitoring process.
   - Step 1 Compare actual results to the operations budget.
   - Step 2 Identify significant variances.
   - Step 3 Determine causes of variances.
   - Step 4 Take corrective action or modify the operations budget.

Quiz Yourself

1. a	3. b	5. d	7. a	9. c

Chapter 12

Key Terms & Concepts Review

1. u	9. gg	17. hh	25. f	33. h	41. q	49. w
3. n	11. xx	19. rr	27. ww	35. ss	43. r	
5. oo	13. c	21. cc	29. vv	37. dd	45. m	
7. i	15. z	23. pp	31. x	39. j	47. e	

Discussion Questions

1. List four business activities that are addressed by capital budgeting techniques.
   - Establishing a business
   - Expanding a business (increase revenues)
   - Increasing efficiency (reduce expenses)
   - Complying with the law

3. Explain the difference between debt and equity financing.
   - With **debt financing**, the investor borrows money and must pay it back with interest within a certain timeframe.
   - With **equity financing**, investors raise money by selling a portion of ownership in the company.

Quiz Yourself

1. c   3. d   5. d   7. b   9. c